No. 628
$8.95

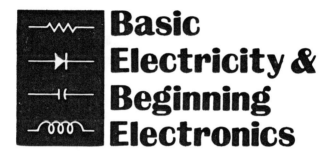
Basic Electricity & Beginning Electronics

By Martin Clifford

TAB BOOKS
Blue Ridge Summit, Pa. 17214

FIRST EDITION

FIRST PRINTING—MARCH 1973
SECOND PRINTING—SEPTEMBER 1974

Copyright © 1973 by TAB BOOKS

Printed in the United States
of America

All rights reserved. Printed in the United States of America. No part of this publication may be reproduced, stored in a retrieval system, or transmitted, in any form or by any means, electronic, mechanical, photocopying, recording, or otherwise, without the prior written permission of the publisher.

Hardbound Edition: International Standard Book No. 0-8306-3628-5

Paperbound Edition: International Standard Book No. 0-8306-2628-X

Library of Congress Card Number: 72-94809

Preface

Electronics is such a glamorous subject it is sometimes difficult to realize it is just an offshoot of the less spectacular but equally practical science of electricity. Not only do electricity and electronics have a common ancestry, but the same fundamentals are used for both subjects. This is advantageous for students, hobbyists, and would-be technicians and engineers for a study of this subject material prepares them for work in both.

It is difficult to state exactly when the science of electricity started, for that all depends on how you define the word. If you consider that electricity began with the curious behavior of amber—a substance that could attract dry bits of straw—then electricity is almost 3,000 years old. What makes this so startling is that it is only within the past few hundred years that the science of electricity began to be pursued vigorously. But when the chase was finally joined, many illustrious scientists devoted their lives to the task, some of them enduring poverty and ridicule for their efforts. The more notable among them are mentioned in this text. Some of them risked their lives, unknowingly, for they were not aware of the tremendous force they were trying to leash. They subjected themselves and others to electrical shocks, playing with electricity in the way a child might play with dynamite. But they did this for the sake of learning, passing their accumulated knowledge on to the next generation.

Fundamentally, electricity and its offshoot, electronics, use very few parts but in tremendous numbers of different combinations. Electricity and electronics involve batteries, generators, coils, capacitors, resistors, tubes and transistors. And these are all basically simple devices. A tube is just a few bits of metal inside an evacuated container. A transistor is a single chemical element with some added impurities. A capacitor is just a pair of metal plates separated by some insulator. A coil is just a length of wire, wound in circular form. And so, with the help of a few chemicals, a few pieces of metal, and some wire we now have electricity and electronics—two sciences on which our technical civilization rests.

Try doing without them!

No autos, no radio sets, no television, no hi-fi, no telephone, no planes, no space exploration, no elevators, no electric lights, no vacuum cleaners, no electric toasters, broilers, heaters, air conditioning. This isn't a complete list—just enough to give you some idea of how dependent we are on electricity and electronics.

And that is the reason for this book—to supply you with an adequate background of technical information and learning so that using this book as a base you can then move forward in electricity, or electronics, or both, much more easily and much more quickly.

Martin Clifford

Contents

1 How It All Started 7
Meet the Electron—Behavior of Electrons—The Electroscope—A Matter of Charge—The Role of Magnetism—Magnetism, Electricity, and Electronics

2 The Long, Long Search For Electricity 26
The Father of Electrical Science—The Friction Machine—The Leyden Jar—The Concept of Plus and Minus—The Earth as a Conductor—The Resistance of Conductors—How Modern Electricity Started—The First Primary Battery—Cells and Batteries—Circuit Symbols—Dry Cells and Wet Cells—Voltage of a Cell—Current of a Cell—The Unit of Current—Internal Resistance of a Cell—Battery Duty—Life of a Battery—Making Battery Connections—More About the Electron

3 How the Electron Was Finally Trapped 55
The Nature of Matter—Other Elements—Compounds—Molecules—Ionization—Bound and Free Electrons—Electron Statistics — Amperes, Milliamperes, Microamperes—The Ampere-Hour—The Meaning of Voltage — Abbreviations — Conductors — Insulators — The Direction of Current—Alternating Voltage and Alternating Current—Opposing Voltages—The Charging Circuit—Combined AC and DC

4 Steps Toward Current Control 78
What Is A Resistor?—Basic Current Control—Fixed Resistors—The Basic Resistor—Codes and Values—Multiples of Resistance—Conversions—The Series Circuit—Ohm's Law—Simplifying Ohm's Law—Examples—Other Forms of Ohm's Law—The Series Circuit and Ohm's Law—IR Drops—Resistors in Parallel—The Series-Parallel Circuit—The Resistor Color Code—Internal Resistance of a Cell—Power—Multiples of the Watt—Power Ratings of Appliances—Wire Table

5 How Voltages Are Generated 135
From a Straight Wire into a Coil—The Induced Voltage—The Basic Generator—Motors vs Generators—The Concept of Inductance—Coils in Series—The Transformer—Producing Voltages—Meters—Motors—Relays

6 How Electricity Is Stored 179
How Electricity is Stored—The Basic Capacitor—Charging the Capacitor—The Dielectric—Action of the Capacitor—Fixed and Variable Capacitors—Capacitor Codes—Units of Capacitance—Factors That Determine Capacitance—Tantalum Capacitors—Working Voltage—Color Codes for Capacitors—Capacitor Leakage—Increasing Capacitance—Decreasing Capacitance—Stray Capacitance—Controlling Charge and Discharge—Capacitors in DC and AC Circuits—Separating AC from DC—The Capacitor as a Bypass Unit—Other Jobs for Capacitors

7 Tubes and Transistors 205
Filament and Cathode—Relative Polarity—The Three-Element Tube—Solid-State Electronics

8 Waves 229
The Inverter—The Sine Waves—Wavelength and Frequency—Phase—Waveform Variation

Index 253

How It All Started

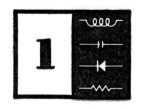

There are some misconceptions about electricity and electronics, one of the most common being that these are relatively new sciences developed within the past few decades. It is true that there have been tremendous applications of both electricity and electronics in the recent past, but consider that the word "electricity" was first used by Sir Thomas Browne in a book published in 1642. To learn about electricity and electronics means we need to go even further back in time, for the ancient Greeks had the word "elektron" and the Romans the Latin word "electrum."

There is an enormous difference between being aware of a natural law and putting that law to work. The knowledge that one substance could attract another was known as far back as about 600 BC. That substance was amber, a brownish translucent fossil resin. It takes a fine polish and is used for making pipe mouthpieces and as a varnish base. But this isn't the most important feature of amber, which originally oozed from trees millions of years ago. Amber has strong electric properties, for when it is rubbed it has the power of attracting certain lightweight materials, such as tiny pieces of paper. When a bit of amber is rubbed vigorously with a clean, dry cloth, it will not only attract small paper fragments, but the paper will fly up toward the amber and cling to it, seemingly in violation of the law of gravity. Amber isn't the only substance to possess this property, for you can rub a glass rod or a rubber comb and get the same results.

It doesn't sound like much. Who could have predicted that our world of radio, television, computers, our entire modern electrical and electronic civilization would start with a bit of amber? When the properties of amber were first realized, paper hadn't been invented as yet, but amber, when rubbed, attracted bits of chaff, the lightweight husks of grain.

MEET THE ELECTRON

It wasn't until the nature of matter was explored and more clearly understood that the mystery of attraction could be explained.

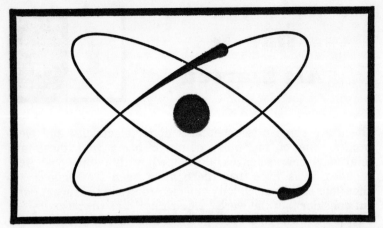

Fig. 1-1. An atom consists of a central portion, the nucleus, with one or more electrons revolving in orbit around it. This is an artist's conception of the helium atom, with two orbital electrons.

Our known universe is made of fundamental building blocks called elements. There aren't too many of them, just a bit more than one hundred, but everything you can see or touch is composed of these elements. Gold is an element. So is silver. And you can include iron. Germanium and silicon, the stuff of which transistors are made, are elements. An element can be a liquid, gas, or solid. Mercury is an element, but it is both liquid and metallic. Hydrogen, the lightest element of all, is a gas. Helium is an element, and so is neon. An element is a substance which cannot be reduced to any more primary form.

Combinations of elements form compounds. Rust is a compound, but we can take rust and separate the rust into its original elements, iron and oxygen. But that is as far as we can go, for iron and oxygen are both elements.

In their turn, elements are made of still smaller building blocks called atoms. Investigating the atom to determine its nature has supplied the answer as to why some substances, such as hard rubber, glass, or amber, will attract bits of paper or straw.

Basically, an atom (Fig. 1-1) consists of a central portion, the nucleus, and extremely tiny particles in orbit around the nucleus called electrons. Electrons, and our ability to control them, form the basis for all modern communications, radio, television, radar, all modern transportation and manufacturing.

A current of electricity consists of the movement of electrons, either through a wire, or some other substance, or through space. Electrons are extremely small and are just about one six-million-millionth of an inch in diameter. In the hydrogen atom, consisting of a nucleus and a single electron revolving around it, the nucleus has a mass about 1,845 times that of the electron. Each electron is a tiny magnet, and it is surrounded by a magnetic field. It is also thought that the electron has angular momentum, spinning or rotating about an axis, somewhat in the manner of the earth turning on its axis.

An electron is matter having a mass that is approximately 0.9 billionth of a billionth of a billionth of a gram, and a gram is equal to 2.2 thousandths of a pound. Actually, the weight of an electron isn't quite as constant as you might imagine for it also depends on the speed with which the electron moves. Light travels in free space at 186,000 miles per second and it is possible to make electrons move at speeds not too far removed from this figure. Mass increases with speed and so the faster the electron travels, the more it weighs.

Electrons are matter and have the properties of matter. They have inertia and remain at rest or in uniform motion in the same straight line or direction, unless acted upon by some external force. This means that electrons oppose being set in motion, but once in motion, resist being stopped. Equally important statistics are that all electrons are alike; they all carry a negative charge (Fig. 1-2); and they all repel each other. And although the Greeks had the word elektron thousands of years ago, it wasn't until 1897 that Sir J. J. Thomson first isolated the electron. Of course electricity was at work long before then. Some twenty years before Thomson, Heinrich Hertz had transmitted radio waves, and Alexander Graham Bell had created his telephone.

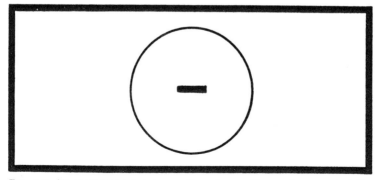

Fig. 1-2. All electrons have a negative charge. No one knows what the electron looks like or if it has any kind of shape at all. It is convenient to represent the electron by a circle and its charge by a minus sign.

Although all electrons are identical, some are associated only with the atoms of the elements. These are bound electrons for they normally do not leave the atom with which they are associated. Other electrons are free electrons: they exist just about everywhere.

BEHAVIOR OF ELECTRONS

Since we can't see electrons, and ordinarily cannot feel them, we are sometimes reminded of their presence, often quite forcefully. Walk across a rug, touch a doorknob, and that small spark you may see and feel is the passage of tremendous quantities of electrons between you and the metal. Sometimes, when you comb your hair, you may find the hair clinging to the comb. If you're having a party, balloons can be made to adhere to walls just by rubbing the balloons lightly. A bolt of lightning is a more violent demonstration of what electrons can do under the right circumstances.

THE ELECTROSCOPE

Since electrons are so extremely small, it is impossible to see them even through the most powerful microscope. Even if we could see them, we wouldn't be able to identify that mysterious something we call an electric charge. However, it is possible to prove the presence of an electric charge, and to learn something of the behavior of such a charge, through the use of very simple experiments.

Take a small section of newspaper and cut it into tiny pieces about ⅛ inch square. Rub a rubber or plastic comb vigorously with a clean dry handkerchief or cloth, and then bring the rubbed end close to the pieces of paper. You may find the paper clinging to the comb. Try other items that won't conduct electricity (insulators) such as a plastic ballpoint pen, your eyeglasses, or a glass swizzle stick used for mixing drinks. Some will barely attract the paper; others will pull up a number of the tiny scraps at one time, with the scraps hanging on to each other. In all instances, if the paper scraps do move, they will move against the force of gravity.

The attracting force is produced by rubbing the insulator (Fig. 1-3) and forcing a tremendous number of electrons to accumulate at the end that is rubbed. Since electrons have the same charge, they repel each other and look for as large a surface area as possible on which to spread out. It is difficult for the electrons to spread through plastic or hard rubber, for these substances are insulators and oppose the movement of the electrons. But the scraps of paper offer more surface area which the electrons can occupy. Try to visualize lines of force (known as electrostatic lines) existing between the rubbed rod and the paper bits. Think of these lines of force as stretched rubber bands attached to the paper and the rod. When the paper flies up to the rod, the electrons move over its surface since a greater area has been supplied.

You can repeat this experiment with a conductor consisting of tiny scraps of copper and a rubbed glass rod. When you rub the glass rod you will once again force electrons to accumulate at the end of the rod. But if you touch the rod to tiny bits of copper, the electrons will move from the glass rod to the copper, and spread through the copper, since this metal is a good conductor. The same series of events will take place as in the case of paper bits except that the copper will not be pulled over to the glass rod. The bits of copper are too heavy, of course, but even if they were light enough, all the electrons demand is more surface area and touching the rod to the copper supplies that demand. Once the electrons have spread through the copper an electrostatic force no longer exists between the copper and the glass rod. We have a state of electron equilibrium with approximately the same distribution of electron quantity on the rubbed rod and the bits of copper.

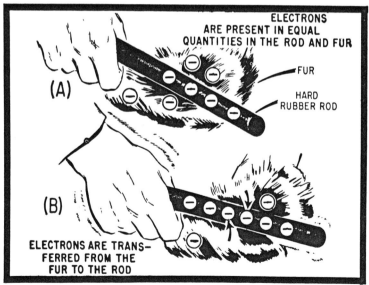

Fig. 1-3. When a hard rubber rod is rubbed with fur, electrons are transferred from the fur to the rod. The rod is then said to be negatively charged.

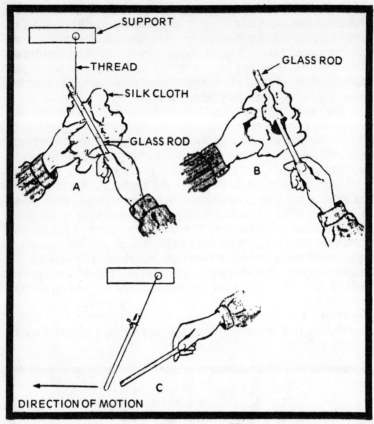

Fig. 1-4. Suspend a glass rod by a thread, and then rub the rod with some silk, as in drawing (A). Rub another glass rod with the same silk (B) and then try to bring the two rods together (C). The suspended rod will keep swinging away and you will be unable to make the two rods touch each other.

There is still another experiment you can do with very little equipment, as shown in Fig. 1-4. The only equipment required consists of two glass rods and some thread. Suspend one of the glass rods by a nylon, silk, or cotton thread. Holding the other glass rod in your hand, rub it thoroughly with the silk. Use the same treatment on the suspended rod. If you will now bring the two rods close together, you will find you will be unable to make the rods touch each other.

Another method of demonstrating electron behavior is through the use of a simple device known as an electroscope, shown in Fig. 1-5. It consists of a glass jar with an insulating stopper, such as cork or rubber. Through the center of the stopper is a metal rod with a pair of gold leaves suspended from the bottom of the rod. A small metal disk is attached to the top of the rod, protruding from the stopper. Gold leaves are used in this experiment because gold, due to its softness and malleability, can be beaten into leaf form. Small sections of lightweight aluminum foil could be used instead of gold. Finally, surrounding the outside bottom portion of the jar is a conducting strip made of some kind of foil.

To use the electroscope, just rub an insulating material such as a glass rod or a hard rubber comb and touch it to the conducting top of the jar. When this happens, the gold leaves will separate and remain that way even after the rod or comb is removed.

The explanation is quite simple. Rubbing a glass rod or comb forces a large number of electrons onto the rod or comb. When the rod is touched to the conducting top of the electroscope, the electrons leave the rod, move through the center conductor to the gold leaves, spreading themselves more or less uniformly across the surface areas of both leaves. The leaves, though, are very close to each other. And since both leaves now have strong concentrations of electrons on their surfaces, they repel each other, bending outward and away from each other. To discharge the electroscope, all that is needed is a conducting material, such as a copper wire, connected between the conducting top of the electroscope and

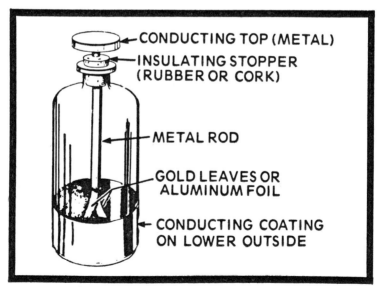

Fig. 1-5. The electroscope can be used to demonstrate the force of repulsion between negatively charged surfaces, and also to supply a rough indication of amount of charge.

the foil coating on the outside. When this connection is made, many of the electrons will travel up through the center conductor to the conductive coating, taking advantage of the larger surface area that has been supplied.

Naturally, some electrons will remain on the gold leaves, but since they will be fewer in number, the leaves will no longer be able to repel each other with the same force and so will assume their original position.

The electroscope is a way of demonstrating that electrons do have a negative charge and that these charges repel each other. The electroscope can also be used to show how effectively a rubbed substance will accumulate electrons. A rubbed glass rod, for example, might cause the gold leaves to separate to a greater extent than a bit of rubbed plastic.

A MATTER OF CHARGE

A glass rod normally has its surface area covered with electrons but since these are distributed fairly uniformly, they exert no electrostatic force or an extremely small force on bits of paper located nearby. The action of rubbing the rod forces a tremendous number of electrons from the fur or cloth being used onto a limited surface area. This action is known as charging. A charged rod can be used to attract paper bits or it can be used to transfer its charge—billions and billions of electrons—to the conducting top of an electroscope. Since all electrons carry a negative charge, technically it would be more accurate to describe the rubbed glass rod that way— negatively charged.

It is only possible to force additional electrons onto a substance, but to take electrons away from it. A substance that has had an electron loss is termed positively charged.

Meaning of Positive and Negative Charges

Positively charged and negatively charged are simply relative terms, for it isn't possible to deprive a substance completely of its electrons. All that can be done is to have a large number of them removed. But since electrons all carry a negative charge, removing them means the substance is not as negatively charged as it was originally. Another way of saying the same thing is to refer to it as positively charged.

As a further example, assume you have two glass rods. Rub one of them with a dry cloth and it will become charged— that is, it will now have more electrons than it had to begin with. Now compare the two rods. The one that was rubbed is negatively charged, but what about the one that wasn't touched? We can say that it is positively charged—that is, it has fewer electrons than the rod that was rubbed.

Now suppose we have a third rod and have somehow managed to make it tremendously negatively charged. The three rods are shown in Fig. 1-6. The new rod, at the right, is the most negatively charged. The one in the center has fewer electrons, and so it is positive with respect to the newly added rod, but is also negative with respect to the completely uncharged rod. And so, the terms positive and negative are purely relative.

The Induced Charge

The action of rubbing a rod forces the accumulation of electrons onto a small area. This charge can be transferred to an instrument such as an electroscope. However, the gold leaves inside the electroscope aren't touched or rubbed. The negative charge is induced—that is, transferred from a previously charged substance.

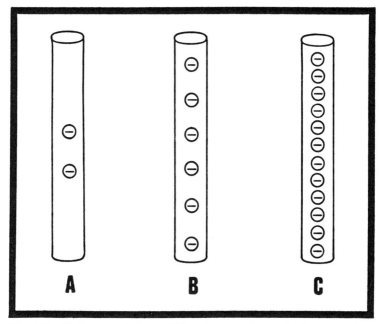

Fig. 1-6. All three rods have different amounts of electrons on their surfaces. The rod at the right (C) has the greatest number, that at the left (A) the fewest. Rod B is positive with respect to rod C (it has fewer electrons than rod C) but negative with respect to rod A (it has more electrons than rod A).

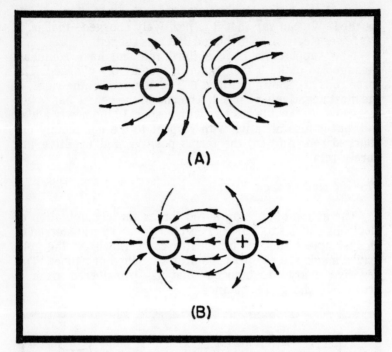

Fig. 1-7. An electric field of force can be considered to exist between two charged bodies. If the charged bodies have the same polarity, the force is one of repulsion. If the charged bodies do not have the same polarity, the force is one of attraction.

Electric Lines of Force

Certain substances, when rubbed, gain electrons; other substances subjected to the same treatment lose electrons. In either case, whether the substance gains or loses electrons, it is said to be charged. If there are more electrons than customary, the substance is negatively charged; and positively charged if there are fewer electrons. While a charged body can be produced by friction, as in the example of rubbing a glass rod with silk or fur, a body can also become charged just by moving. If you walk across a rug, your body becomes charged because its surface has accumulated electrons by the shuffling of your shoes across the rug. You cannot feel these billions of electrons that are now more or less uniformly scattered over your skin area. But if you touch a conductor, such as a doorknob, the electrons will move away from you even more rapidly than they arrived. In a quiet, dark room you may both see and hear the spark of electricity—the result of the movement of electrons—from your finger to the knob.

A gasoline truck rumbling along a road becomes charged, and so to prevent an accidental spark, it is equipped with discharge devices. Tiny metal rods discharge your car when it drives up to a toll booth, otherwise the toll collector would constantly be shocked by all the electric discharges from automobile drivers.

A cloud moving through the sky gathers a charge. When it has built up sufficiently it discharges violently in what we see as a lightning display.

Radiating from each charge are electric lines of force, also known as electrostatic lines. These lines are invisible, yet they exist and can be put to work. If two plates, equally charged by extra billions of electrons on their surfaces, are brought near each other, the repulsion between the plates can be shown as in Fig. 1-7 (drawing A). If, however, one of the plates has many more electrons than the other, the plate with the electron surplus is negatively charged (since each electron is negative) while the other plate is positively charged.

Coulomb's Law of Charges

If two substances are charged, both may be negative or one may be positive and the other negative, depending on the electron distribution on the surfaces of the substances. In either instance, there is a force of attraction or else a force of repulsion between them. The amount of force depends on two factors: the extent to which the bodies are charged and the distance between them. The greater the charge on one or both of the substances, the greater the amount of force that exists between them, that is, the greater will be the effort made by the bodies to move together or to move away. The force will be greater if the bodies are closer together than if they are separated.

This relationship between the amount of charge, distance, and the amount of electrostatic force was originally discovered by Charles A. Coulomb and is now known as Coulomb's law: "Charged bodies attract or repel each other with a force that is directly proportional to the product of their charges and inversely proportional to the square of the distance between them."

This law can be stated much more concisely in the shape of a very simple formula, in this way:

$$\text{force} = \frac{Q1 \times Q2}{d^2}$$

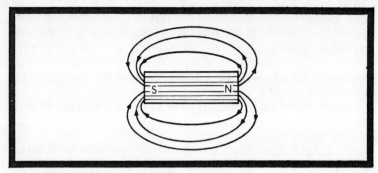

Fig. 1-8. Magnetic lines of force are invisible but are drawn here to show what they might look like. Conventionally, it is assumed that magnetic lines of force start at the north pole, continue through the space around the magnet to the south pole, and then exist inside the magnet between the two poles.

Q1 and Q2 are the charges; d is the distance. Multiply Q1 by Q2 and then divide by the distance squared. The result is the amount of force existing between the two bodies.

Static vs Dynamic Electricity

Electrons can be collected on a surface, such as a glass or rubber rod, simply by using friction. Electrons are transferred from the fur to the rod, but aside from this very limited movement, the electrons remain on the rod until they are given an opportunity to get off. Since the electrons remain motionless for the most part, this kind of electricity is known as static electricity. However, as you will see in later chapters, we are more concerned with current or dynamic electricity—electrons which are ordinarily in motion through a wire, or space, or through some conducting liquid.

THE ROLE OF MAGNETISM

Another of the amazing properties of electrons is that each electron in motion behaves like a magnet. However, electrons aren't the only kinds of matter that show this property. The earth is a huge magnet, a most useful fact for navigators.

The word magnetism stems from the Greek city-state Magnesia in Asia Minor, a region rich in magnetic ores. These ores, known as magnetite, were also called lodestone or leading stone, since the stone, when freely suspended, would align itself in a north-south direction. Magnetite, or lodestone, is iron oxide. But while the characteristics of magnetite were known more than 2,000 years ago, it wasn't used as a compass until about the 12th century, even though the Chinese claim to have invented the compass as early as 2637 BC.

Magnetite has unusual properties. It can attract small bits of iron or steel and can also be used as a compass. If a steel needle is rubbed along a section of magnetite, the needle also becomes a magnet. An early form of compass consisted of a needle which had been magnetized, placed on a splinter of wood, floating in water. A magnet can also be made by rubbing a small length of steel with magnetite.

Magnetic Poles

A steel bar that has been magnetized looks exactly the same as a steel bar that hasn't been touched. When a bar is magnetized it does not change its shape, its color, its weight, the way it looks or feels. And yet, when a steel bar is magnetized, something must happen to give it the unusual ability of attracting and holding bits of iron. One theory is that the act of magnetization aligns all the molecules of the bar so that the magnetic forces of all the molecules of steel are cumulative.

The ends of a magnet are referred to as its poles (Fig. 1-8). One end is the north pole; the other end the south pole. When a magnet in the shape of a bar is covered with a sheet of paper sprinkled with iron dust, the dust arranges itself in the form of a pattern, somewhat like that shown in Fig. 1-9. The lines which appear in the space between the poles are called magnetic lines of force or flux. Each line is continuous, unbroken, and the magnetic lines of a magnet cannot cross each other. Starting at the north pole, you can follow a single line

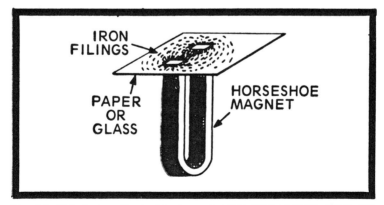

Fig. 1-9. This is a method of seeing magnetic lines of force. When the paper or glass is tapped gently, the filings will arrange themselves in the form of lines, representing the magnetic lines of flux between the poles.

outside the magnet until you reach the south pole. The line of force then continues through the body of the magnet, up to the north pole. In this example, the line of flux is steady and does not move. Nothing travels along the flux line. It is simply a line of force.

Flux Density

A single line of flux is sometimes called a maxwell, in honor of James Clark Maxwell, a Scottish physicist, who advanced the theory, in 1865, that electrical impulses could travel through space with the speed of light. What Maxwell did, in short, was to predict, mathematically, the possible existence of radio waves.

The maxwell, though, is a small unit, and so an alternative unit, the weber, is sometimes used. One weber equals 100,000,000 maxwells, or one hundred million lines of magnetic flux. Weber was a noted German physicist who did experimental work in telegraphy.

Flux density, as its name implies, is the number of lines of flux, or maxwells, for a particular area. Flux density can be expressed as the number of lines per square centimeter or lines per square inch. The line per square centimeter is called the gauss, named after Karl Friedrich Gauss. Gauss, together with Wilhelm Eduard Weber, worked in a magnetic observatory in 1833 constructed almost completely of non-magnetic materials and made observations of the earth's magnetism.

Number of Poles

Every magnet that has a pair of open ends has two poles: a north and a south pole. If a magnet if cut in half, each of the sliced sections will behave as individual magnets and each will have its own north and south poles. However, if a pair of magnets are joined, so that the north pole of one butts against the south pole of the other, the two joined pieces will still have only two poles, and there will be no poles evident at the joining area.

The Magnetic Field

The total number of lines surrounding a magnet is known as a field. A magnetic field may be strong—that is, there may be many lines per square inch, or the magnetic field may be weak, with relatively few lines.

The earth, like all magnets, has a pair of poles; a north magnetic pole and a south magnetic pole. These poles are not the same as the geographic North Pole and the South Pole. The geographic North and South poles represent the points of the earth's rotational axis. The earth's magnetic poles are the two spots on the earth's surface toward which a compass needle will point, and are located some distance away from the geographic North and South poles.

Laws of Magnetism

A compass is nothing more than a tiny magnet, pivoted so it can swing freely at its center point. If a pair of compasses are brought adjacent to each other the north pole of one compass needle and the south pole of the other will constantly point to each other and try to be as close as possible. Opposite poles of magnets attract each other. However, if you try to bring two magnets together so that the north pole of each magnet faces the other, you will see or feel the magnets pushing away (Fig. 1-10). The same effect takes place when a pair of south poles are placed adjacent; they repel. And so the rules of magnetism are quite easy: unlike magnetic poles attract each other, and similar magnetic poles repel each other.

Magnetic Materials

There are a limited number of materials from which magnets can be made. These include iron, steel, and various

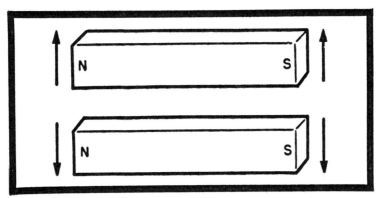

Fig. 1-10. If similar poles of magnets are made to face each other, as in this illustration, the magnets will repel each other. If the magnets are strong enough, the upper magnet can be made to float in space above the lower magnet.

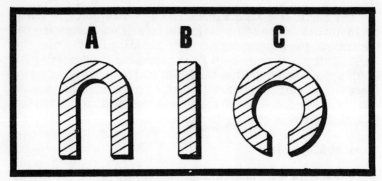

Fig. 1-11. Three commonly used shapes of magnets. The horseshoe magnet (A); the bar magnet (B) and the ring magnet (C).

alloys or mixtures consisting of aluminum, nickel and cobalt, or nickel and iron, or nickel and molybdenum, or an alloy of nickel, copper, and molybdenum. Generally, a soft metal will not hold its magnetism for long. The word soft is used here in the sense that the metal is malleable—it can be shaped or machined quite easily. Hard metals, such as steel, retain magnetism for much longer periods.

Paramagnetic and Diamagnetic Materials

Substances that are attracted by a magnet, such as iron, steel, nickel and cobalt, are called magnetic. While most other substances are nonmagnetic, some are slightly attracted while others are just slightly repelled. Those that are faintly attracted are called paramagnetic, while those that are weakly repelled are known as diamagnetic. Aluminum, platinum, sodium, potassium and oxygen are representative paramagnetic substances. Bismuth, copper, gold, antimony, mercury, carbon, water, and hydrogen are part of the list of diamagnetic substances.

Magnetic Shielding

Some substances are magnetic, that is, they can be magnetized; but most are not. Glass, wood, paper, plastic materials, rubber, are all examples of nonmagnetic materials. If a sheet of glass is placed on a magnet, the lines of flux of the magnet will pass right through the glass and the magnetic lines will not be affected in any way. The glass face on a compass does not interfere with the reaction between the earth's magnetic lines and the magnetic lines around the compass.

However, if a magnet is covered by a sheet of iron or steel, the magnetic lines of force of the compass will pass through the iron or steel in preference to air. Iron and steel have a higher permeability than air, just another way of saying that they offer an easier path for magnetic lines of force than air. Thus, if a magnet is completely surrounded by a box made of iron, none of the lines of magnetic force will be detected outside the box. The lines of force of the magnet will extend from the magnet to the box, through the box and back to the south pole of the magnet, and then through the body of the magnet back to the north pole. A similar box, made of cardboard, wood, glass or paper, will have no effect whatsoever on the magnetic lines.

Shapes of Magnets

A magnet can be made into any shape, but the three most commonly used are the bar magnet, the horseshoe magnet and the ring magnet, all shown in Fig. 1-11. The shape of a magnet is decided by its commercial application. Thus, in meters used for measuring voltage and current, the magnets are horseshoe or ring types. The advantage of bending a magnet into a horseshoe or ring is that the two poles of the magnet are much closer to each other. The magnetic lines of force no longer need to pass through the relatively greater distance required by the bar magnet. Instead, the magnetic lines of force are concentrated in the small region between poles of the magnet. Air has quite a bit of opposition to magnetic lines, a property known as reluctance. This reluctance is decreased as the space between the poles of a magnet is decreased. As a result, there can be a greater concentration of lines between the magnet's poles—that is, there will be more lines, or a higher flux density.

The Basic Compass

A compass is simply a tiny, lightweight permanent magnet—a tiny sliver of steel that has been magnetized—supported by a centrally located pivot. Any bar magnet can be made into a compass, as shown in Fig. 1-12. The bar magnet is suspended by a thread wrapped around its center, so arranged that the bar magnet will be horizontal and will be free to rotate. When this happens one end of the magnet will point toward the earth's north magnetic pole. This end of the magnet is actually its south pole, since unlike poles attract each other. However, it may be marked with the letter N, not

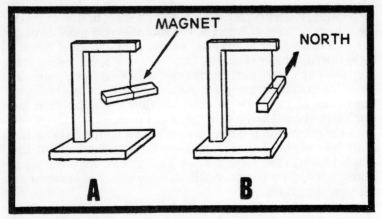

Fig. 1-12. A suspended magnet behaves like a compass. The magnet will align itself so that one end of the magnet points north.

to indicate that it is a north pole, for it is not, but rather that it is a **north seeking** pole. Generally, however, the magnet used in a compass isn't marked with the letters N or S. Instead, a geographic scale, indicating north, south, east and west, and intermediate points, is placed below the movable compass magnet.

The Basic PM Motor

A motor is just a device that takes advantage of the fact that similar magnetic poles repel and unlike magnetic poles attract. It's easy to build a motor using a pair of magnets, or a compass and a magnet. Suspend a bar magnet with a string around its center, so that the magnet is free to rotate. Bring another magnet close to the suspended magnet so that the string supported magnet is repelled. You can now force this magnet to rotate, but what you have, in effect, is a motor. The difference between this motor and one that can actually be

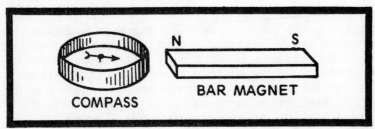

Fig. 1-13. In its present position, the north pole of the bar magnet will attract the south pole of the compass.

used as a motor is in construction and design, but not in principle. You can also use a compass and a magnet (Fig. 1-13) to perform the same experiment. Put the compass on a table and bring any kind of magnet near it. With the help of the magnet you will be able to make the compass pointer spin on its pivot. Again, this is a demonstration of the way a motor works.

MAGNETISM, ELECTRICITY, AND ELECTRONICS

It does seem as though we have very little to work with. A rubber or glass rod which is rubbed and which can then attract small pieces of paper, and a magnet which can only attract a very limited number of substances, such as iron or steel. It doesn't look like a hopeful beginning at all, and yet this is all we need as a start in electricity and electronics. However, this first chapter is just an introduction, for while the attraction of substances for each other may be interesting, it is quite another matter to put such knowledge to work. To be able to do so, it's necessary to learn more about electric currents and how they can be produced. That's in the next chapter.

2 The Long, Long Search For Electricity

When a nonmetallic substance such as amber is rubbed it acquires the ability to attract bits of paper or straw. It does so because the rubbing process transfers electrons to the amber. We can consider electrons as the center point for lines of force—electrostatic lines reaching out in all directions. The force of attraction between a charged body, such as amber or a glass rod, and the bits of paper can be measured. And, although the electrostatic lines of force cannot be seen, they are put to work in various electronic devices.

The difference between a glass rod that has been rubbed and a magnet is that the magnet does not need to be rubbed with a cloth. Further, the magnet is surrounded by a different kind of lines of force—magnetic, or flux. And, as in the case of electrostatic lines, magnetic flux lines cannot be seen, but they can also be put to work.

And so our world of electricity rests on these two very strange ways on which certain substances behave. However, in the time of Thales of Miletus (about 640 to 548 BC) the explanation given for the attractive powers of amber and lodestone was that these materials had a soul. In the hundreds and hundreds of years that followed various other explanations were supplied. Theophrastus of Eresos in Lesbos (372 - 287 BC) wrote: "Amber is a stone. It is dug out of the earth in Liguria and has a power of attraction. It is said to attract not only straws and small pieces of sticks, but even copper and iron if they are beaten into thin plates." The theories of the ancient Greeks about amber, magnetite, and various natural displays such as lightning, and the aurora borealis, are interesting but not instructive for they saw no relationship between amber, magnetite, and lightning. Thus, Plutarch (about 46 to 120 AD) explained that the reason it was necessary to rub amber was to open the pores of the amber to permit the surrounding air to rush in to fill the empty spaces, thus dragging along light particles of straw.

The ancient Romans speculated even less than the Greeks on the nature of amber and lodestone. The Romans were soldiers, excelled in construction engineering and were known as superb orators, but they weren't mathematicians or scientists. Following the Romans it took about 1500 years before any serious efforts were made to learn more about the properties of amber and lodestone. All that happened in the intervening centuries was that amber was cut, polished, and shaped into jewelry.

THE FATHER OF ELECTRICAL SCIENCE

The trouble with the Greeks and Romans in the matter of amber and lodestone was that they believed in discussion, not experimentation. As far as electricity is concerned, it wasn't until around 1600 that the concept of learning about electricity by working with it took hold. In that year, William Gilbert, a physician and a physicist, published "De Magnete," not devoted to electricity, but containing a chapter on that subject. Gilbert coined the word "electrica" and also learned that amber wasn't the only substance that had powers of attraction when rubbed. He included glass, sulphur, alum, rock salt, precious stones, and rock crystal. To be able to detect the powers of attraction of these materials he invented a sort of electroscope he called a versorium. The important feature of Gilbert's work is the emphasis he placed on experimentation. It was this aspect that was more important than increasing the list of substances that had powers of attraction when rubbed. Gilbert also tried to find the relationship between static and magnetic forces—that is, a common denominator for substances that behaved like amber, and lodestone.

THE FRICTION MACHINE

Despite Gilbert's achievements, it wasn't until about 75 years later that the idea of learning about electricity by direct personal involvement took place. In 1672, Otto von Guericke made a globe of sulphur which was so mounted that it could be rotated. With the friction of his hand resting against the revolving globe, von Guericke was able to produce electrical sparks. This could probably be termed the first electric generator. It was impractical, it could not produce a sustained electric current, yet it was the first man-made electricity that could be seen. Unfortunately, what was not known then was that electrostatics, or the production of electricity by friction, was a secondary road to electricity. It wasn't until moving electric currents were produced that electrical experimentation made progress toward the kind of electricity we know today. But to give von Guericke credit, he did illustrate the great importance of doing electrical experiments and he was also the first to discover that there was also a force of repulsion, not just forces of attraction.

Referring to William Gilbert again for a moment, his book was primarily on magnetism, and was written in Latin. The first book on electricity written in English was by Robert Boyle, in 1675. That isn't so long ago when you consider that i' was almost 200 years after the discovery of America by Columbus. Also consider that we now have computers, electric lighting, electric heating, radio, television, radar, motors, generators, electronic calculators, electric cooking, plus hundreds and hundreds of other electric and electronic devices, just about 300 years after the first book on electricity. One of Boyle's great contributions was his continuing emphasis on the need for experimental work.

THE LEYDEN JAR

In the years following von Guericke, various types of friction machines were developed for the production of electrostatic electricity. Sir Isaac Newton conceived the idea of using a glass globe instead of one made of sulphur, and then, about 1742, Andreas Gordon, a professor of philosophy, used a glass cylinder; but subsequently, other investigators replaced the cylinder with a glass disc. These friction machines became more and more elaborate. Numerous attempts were made to carry the electricity generated away from these friction generators. The trouble, of course, was that these investigators were using the wrong kind of generators. But these glass-disc friction machines did produce enormous sparks and the spectacular results, even though impractical, did encourage other investigators. The effort to lead generated electricity away from the friction machines led Stephen Gray to the conclusion that some materials were conductors of electricity, while others were nonconductors (insulators).

Electrical experimentation then began to follow three directions: 1) the generation of electricity through the design of various types of friction machines; 2) the transmission of electricity away from the generating device; and 3) the storage of electricity. This is the elementary foundation of modern electricity and so we can see that these early investigators had a general idea of the direction their investigations should follow.

The concept of storing or bottling electricity was made by Ewald George von Kleist, about 1745, with an invention of the device made independently by Professor van Musschenbroek of the University of Leyden in Holland. The unit for storing electricity, called a Leyden jar, was originally a glass bottle containing water. One wire connecting to the friction generator was placed in the water. The operator of the generator then put his hands around the outside of the bottle. Actually, the Leyden jar was the first electric capacitor, a component now used in a tremendous number of electric and electronic applications. In its fundamental form, a capacitor consists of a layer of insulating material sandwiched between two metallic plates. In the original Leyden jar, the water took the place of one of the metallic plates, while the hand of the operator acted as the other plate. The glass of the jar was the insulating material between these two conductors. In later developments, the water was replaced with iron filings (a much better conductor than the water) while sheet lead was used outside the glass (a much better conductor than the hand of the operator). Finally, one investigator realized that the shape of the bottle had nothing to do with the ability of the Leyden jar to store an electric charge and so designed a "Leyden jar" consisting of a flat plate of glass with a tinfoil coating on both sides. In this way the Leyden jar gave birth to the first true capacitor.

THE CONCEPT OF PLUS AND MINUS

Benjamin Franklin, probably better known for his famous kite experiment, made some interesting discoveries while experimenting with Leyden jars. He thought that the charges on the metal plates on either side of the glass were not the same, and that they were equal in amount. He called the charges on one plate "positive," and on the other plate, "negative." He also learned that the charge could be increased by decreasing the thickness of the glass and by increasing its area. These fundamental facts are used today in the manufacturer of capacitors.

THE EARTH AS A CONDUCTOR

With the help of the Leyden jar, an electric charge could be stored and carried from one place to another. The trouble was that once the Leyden jar was discharged, it could supply no further electricity until it was charged once again by connecting it to a friction type electric generator. It was around this time, about 1747, that the idea of a closed electric circuit was first developed. That is, investigators now realized that to have electricity move from the friction generator to some distant spot various elements were needed: 1) a device for producing electricity; 2) a conductor leading from the generator to the distant point; 3) a conductor from the distant point back to the generator; and 4) at the distant point some sort of device which would accept the delivery of the electric charge. And this is a rather precise description of a closed circuit.

The idea that the earth could be used as part of this closed circuit occurred to Dr. William Watson. In 1747 he discharged a Leyden jar (which had previously been charged by a friction generator) through a wire which crossed over Westminster Bridge in London, a structure located over the river Thames. The return path was the river itself. Subsequently, the experiment was repeated using the earth as a return path. Today, a "ground return" is used as a matter of course in electrical installations.

When the concept of a closed circuit became known, the race was on to make the circuit longer and longer. In 1748, Dr. Watson was able to discharge a Leyden jar through more than two miles of wire. Again, the problem confronting Dr. Watson, and other investigators, was that their current source was a one-time affair. As soon as the Leyden jar was discharged, it had to be returned to a friction machine for charging again.

THE RESISTANCE OF CONDUCTORS

Investigators of electric circuits learned that not all conductors worked equally well, and that under certain circumstances, materials which are normally insulators can be made into conductors. Joseph Priestley (1767) demonstrated that glass, if made red-hot, becomes a conductor. Henry Cavendish, in 1776, made extensive experiments on the resistance of conductors. He was the first to indicate that iron wire was far superior to rainwater (pure water) for conducting electricity, but that the addition of salt to water increased its conductivity tremendously. Other investigators learned that a very fine wire for transmitting the charge of a Leyden jar became hot enough to melt. Although it doesn't seem too consequential in the light of what we know today, all of these experiments and the data they supplied began to form the basis of a new science—electricity. Electric generators of considerable size had been constructed and, while these generators do not form the basis of the generators we have today, they did supply an electrical energy source for experimenters. Electrical storage devices, known today as capacitors, and then as Leyden jars, were improved. Connecting wires which we call transmission lines today were of all sizes, shapes, and lengths. Their conducting properties were examined and recorded. Again, it was a foundation on which a science of electricity could be built. Unfortunately, there was one component in this entire chain which kept experimenters from pursuing a more appropriate course of action. That component was the friction generator. All the other components of the closed electric circuit are used today: the conductor connecting the generator to some remote point; the device at the remote point for accepting the electric charge (today we call that device a load); the earth as a ground return for the passage of a charge. But now the friction generator is found only in museums or in school laboratories for demonstrating the production of an electric charge by friction.

HOW MODERN ELECTRICITY STARTED

If a symbol were ever needed to represent the science of electricity, perhaps the best one to use would be that of a frog. Stranger things may have happened, but surely ranking at the top of all unusual events would be the experiments made by Alessandro Volta on the nerves of a frog. Using two different metals, Volta noted that these caused contractions when put in contact with a frog's nerve. In 1796 Volta went a step further and prepared large "coins" of copper and zinc. He brought the discs together so that the faces of the discs were in contact. He then separated them, and, using an electroscope, discovered that each of the discs had a very small electrical charge. The charge on the zinc was positive; that on the copper was negative. He then tried using any number of discs, with the collection known as a pile. What Volta had invented was the battery and it was with this invention that we have the missing link needed to replace the friction generator.

Volta improved on his original experiment by using discs of different metals separated by wet cardboard. It is true that the voltaic (named after the man) pile produced a very weak amount of electricity compared with the formidable spark that could be produced by a friction generator, but Volta's unit had one distinct advantage: It did not need to be turned, or operated, or rubbed.

It was also superior to the Leyden jar in the sense that it did not need to be recharged. The voltaic pile combined the advantages of the friction generator and the Leyden jar in one unit. The friction generator produced electricity and the Leyden jar could be used to carry that electricity, in the form

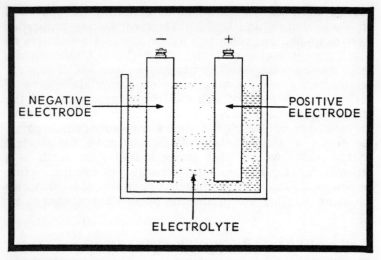

Fig. 2-1. A basic cell consists of two different conductors (electrodes) in a conducting solution (electrolyte). The electrode having more electrons is marked minus; the other is plus.

of a charge, to some distant point. The voltaic pile not only generated electricity, but, like the Leyden jar, could also be easily transported from one place to another.

THE FIRST PRIMARY BATTERY

The voltaic pile was crude, inefficient, produced a very small amount of electricity, but it was the first battery and even more important, it was a start. Fortunately, improvements were not long in coming. Shortly after Volta announced his invention, Dr. William Cruickshank made a long box which contained metal partitions. Each of these was made of a flat section of copper. He also included flat sections of zinc. The two different metals were then supported in grooves which were cut into the inside the wooden box. Instead of using wet cardboard, Dr. Cruickshank tried different solutions of salt, and then used acid.

This battery was the forerunner of those we have today, which use the same basic principle. The battery consists of two different conductors, generally metals, placed in an acidic or basic solution. Each of the conductors is called an electrode, while the liquid between them is referred to as electrolyte (Fig. 2-1). In a modern battery, for example, the electrolyte is a dilute solution of sulphuric acid (Fig. 2-2) while one electrode is pure sponge lead, and the other is lead peroxide.

Various other investigators interested in electricity followed Volta and Cruickshank, working with different metals for the electrode and using different kinds of electrolyte. And they did manage to produce a variety of batteries. The important fact, though, is that the battery had been invented.

CELLS AND BATTERIES

Although the terms "cells" and "battery" are often used interchangeably, a cell is a single unit, while a battery consists of two or more cells. Cells may be connected in a variety of ways, but no matter how they are wired to each other, any combination of two or more in a single unit is referred to as a battery. Your car has perhaps six 2-volt cells, but since they work as a team, the combination of the cells is a battery.

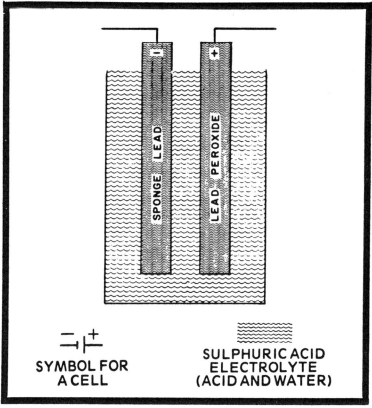

Fig. 2-2. Basic construction of the lead-acid cell, which changes chemical energy to electrical energy.

Types of Cells

Since the time of Volta a large number of cells have been devised using just about every possible combination of electrode and electrolyte. In the Daniell cell, zinc and copper are used as the electrodes but there are also two different electrolytes. The copper electrode is put into a saturated solution of copper sulphate while the zinc is in a dilute solution of sulphuric acid. The two electrodes and their individual electrolytes are separated by porcelain.

The Grove cell uses zinc and platinum as the electrodes with the zinc in sulphuric acid and the platinum in nitric acid. Another electrical investigator, von Bunsen, used carbon instead of platinum as an electrode.

How a Cell Works

Fig. 2-2 shows a simple cell consisting of two electrodes, one made of a strip of copper and the other of zinc. The two electrodes are immersed in a solution of water and sulphuric acid.

When the electrodes are placed in the acidic solution, the acid removes electrons from the copper strip. Since electrons carry a negative charge, the copper strip (having lost electrons) is no longer neutral, but has become positive. It is important to note that the copper strip isn't completely deprived of all electrons and will still have tremendous quantities of them. However, the loss of electrons or negative charges means that the copper plate is no longer as negative as it was originally (is positive).

The electrons move through the acid electrolyte and collect on the zinc strip. With an accumulation of electrons, the zinc strip is also no longer neutral, and may be regarded as negative.

Relative Polarity

We now have an unequal distribution of electrons, with many more electrons on the zinc strip than there are on the copper. It is for this reason that the zinc is negative with respect to the copper. We can also say that the copper is now positive with respect to the zinc.

Now imagine that a second copper strip is inserted into the electrolyte, in any convenient place. This copper strip will also lose electrons to the zinc. Suppose that this second strip loses as many electrons as the first copper strip. Both copper strips will then be positive with respect to the zinc. But what about the relationship of the first copper strip to the second? They will be neutral with respect to each other since both will have the same number of electrons. Thus, a substance isn't positive or negative by itself, but always with respect to something else. In the example given, there is a difference in the number of electrons on the zinc compared with the number of electrons on the copper strips. This difference in electron quantity is what is important.

Electromotive Force

When a glass rod is rubbed with a bit of fur, electrons are transferred from the fur to the rod. The fur loses electrons, hence is more positive than it was before. The glass rod gains electrons, therefore is more negative. If the charged rod is brought near a substance that has far fewer electrons—such as a doorknob—electrons may jump from the rod to the doorknob. Whether or not such a transfer will take place will depend on the distance between the rod and the knob and the number of electrons gathered on the glass rod.

Note how similar this is to the action in the cell shown in Fig. 2-2. Electrons are stripped from the copper and transferred to the zinc. The transfer of electrons, however, takes place by chemical action, not by rubbing. Thus, we now know two ways of producing a charged body: 1) by friction and 2) by chemical means.

An electrical pressure now exists between the two electrodes shown in Fig. 2-2. This electrical pressure is also known as electromotive force (abbreviated as emf) or voltage. Voltage is also called potential difference. Rubbing a rod with a cloth or fur produces a voltage; so does immersing metals in acid. The basic unit of electromotive force is called the volt, in honor of the man who invented the cell.

Positive and Negative Plates

Since the copper strip in the cell of Fig. 2-3 is positive, it is now called the positive plate. The zinc strip, having gained electrons, is referred to as the negative plate.

Putting the Cell to Work

The basic lead-acid cell shown in Fig. 2-2 isn't a complete circuit, but it can be connected into one as shown in Fig. 2-4. In this drawing a pair of copper wires (or similar conductors)

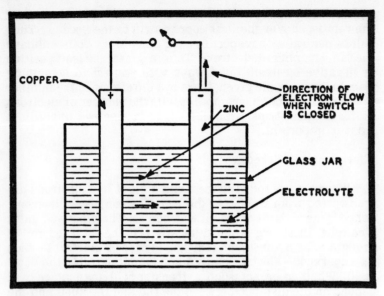

Fig. 2-3. The copper-zinc cell. Because of the action of the acid, electrons are removed from the copper electrode, making it positive. The zinc gains electrons and so is negatively charged. This electron difference constitutes a voltage.

have been connected to the terminals of the electrodes. At the end of these conductors there is a small electric light bulb.

The electrons gathered on the sponge lead strip aren't there by choice but have been forced onto that electrode by the chemical action of the dilute sulphuric acid. Since electrons repel each other, they will try to leave, given the opportunity to do so. And so, in Fig. 2-4, the electrons that originally came from the lead peroxide strip will now move from the pure sponge lead electrode through the connecting wire up to the lamp. The lamp contains a fine wire known as a filament. Passing through the filament, the electrons will return to the lead peroxide strip. No electron will actually travel the circuit; there are too many others in the way. But as each electron tries to "go," it excites another in its path. The result is a wavelike action resembling a tumbling string of dominoes. However, it is convenient to speak of electrons as travelers, so we do so.

The action doesn't stop at this point. The electrons, now back at their starting area, the lead peroxide strip, are once again removed by the acid electrolyte and transferred to the sponge lead. But here they are still being given a chance to escape and so they will continue to move along the conductor, to and through the filament of the lamp, back to the lead peroxide strip. At this time the electrons will once again be taken away by the acid, and so the entire process is a continuous, repetitive one.

Note what has happened, though. In the case of a rod which has been rubbed, there is a single transfer or movement of electrons from the rod to some other substance. Once that happens the rod is no longer charged and the entire action is completed. This is not the situation in the case of the cell. What

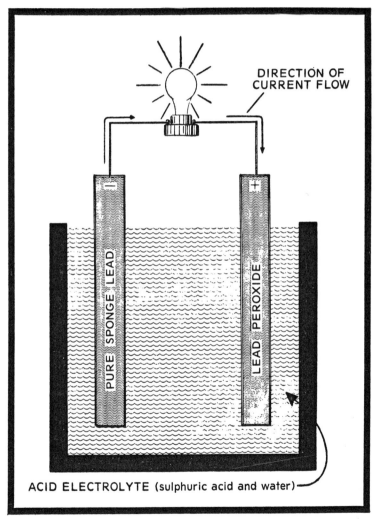

Fig. 2-4. The cell circuit is completed when electrons are given an opportunity of moving from the negative to the positive terminal external to the cell.

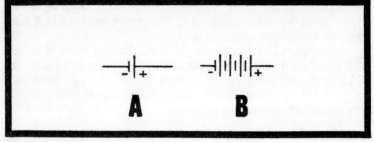

Fig. 2-5. Drawing A shows the symbol for a cell; B is the symbol for a battery. The short line represents the negative terminal; the longer indicates positive.

we have here is a continuous movement of electrons. This "flow" constitutes an electric current and because it moves continuously is called dynamic. The movement of charges from a rubbed rod is called static, because, for the most part, when the electrons gather on the rubbed surface of the rod, they remain there.

The Load

The arrangement of the cell, the connecting wires, and the lamp is called a closed circuit because the electrons are supplied with a complete path. This means the electrons can always get back to their starting point. The lamp is known as a load because the electrons can be made to perform useful work when they flow through it. The wire inside the lamp is especially designed to become white-hot when an electric current flows. The heat is incidental; the light is what we want. But the wire would burn up instantly if exposed to air, so the filament is enclosed in a transparent airtight chamber.

The circuit of Fig. 2-4 is closed whether the lamp is located just a few inches away, a few feet, or a few feet, or a few miles from the cell. The toaster you use in your home, the various electric lights, your radio or your television set, are all loads. When you connect them to a power outlet you complete a circuit and give an electric current an opportunity to flow through them. It is this electric current that does the work: heat in the case of a toaster; light in the case of electric bulbs; and sound in the case of the radio receiver.

Life of a Cell

The cell of Fig. 2-4 will not supply current indefinitely. In time the electrodes are destroyed by the acid and must be replaced. While this is easy enough to do in a laboratory, it would be a nuisance in a practical application. And so, modern cells are capable of being recharged—that is, restored to their original condition.

CIRCUIT SYMBOLS

A circuit drawing using the method shown in Fig. 2-4 is called a pictorial. While this type of circuit drawing is easy to understand it isn't so easy to draw and is somewhat time consuming. A simpler method is to use designated symbols to represent cells, connecting conductors, and a load. The symbol for a cell consists of a pair of parallel lines, one a little longer than the other (Fig. 2-5). The longer of the two lines represents the positive electrode; the shorter, the negative electrode. The mathematic symbols for plus and minus (+ −) are often placed adjacent to the two lines. The symbol may be drawn in any position, vertical or horizontal. A diagram using various symbols is called a schematic.

The symbol for a conductor is a line which may be straight, curved, or bent at some angle. The lamp also has its own symbol, as shown in Fig. 2-5. A complete pictorial electric or electronic diagram appears in Fig. 2-7A and corresponds to the schematic diagram drawn in Fig. 2-7B.

As explained earlier, the lamp is also called a load (Fig. 2-8) for it is operated by current taken from the cell. A motor, or a toaster, or a radio would also come under the general heading of load. Sometimes, in a circuit, it is convenient to indicate a load, without specifying the type of load. In that case the load symbol is used, as shown in Fig. 2-9.

DRY CELLS AND WET CELLS

When a cell is connected to a load such as a lamp, it is not uncommon for the lamp to remain lighted for a considerable time. However, there are loads which are operated quite intermittently. As an example, consider an electric doorbell.

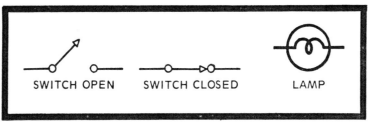

Fig. 2-6. Various symbols used in electrical and electronic circuits. Connecting lines always represent conductors.

39

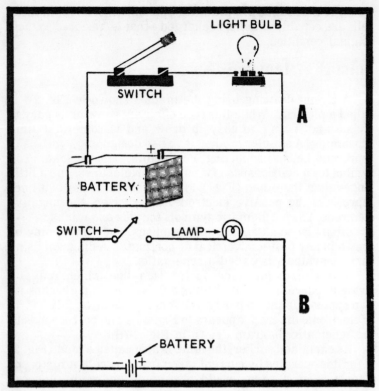

Fig. 2-7. Drawing A is a pictorial of a circuit, while drawing B is a schematic which uses symbols for the same arrangement.

The bell is the load but it rings for just a short time, and infrequently at that. A flashlight is another example of this kind of short-time operation.

A cell often used for loads of this type is known as a dry cell. As shown in Fig. 2-10, it consists of two electrodes: a carbon rod extending almost all the way down through the center of the cell and an outer shell made of zinc which also serves as the container. The electrolyte isn't a liquid but consists of a mixture of granulated carbon and manganese dioxide.

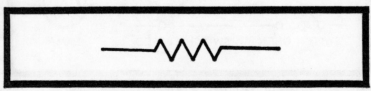

Fig. 2-8. The symbol for a load is a zigzag line.

A cell using a liquid electrolyte is called a wet cell. Ordinarily, dry cells cannot be recharged. The electrolyte and zinc change chemically, but this change isn't reversible. Wet cells such as the lead-acid type can be recharged.

Dry Cells

The dry, or primary, cell, isn't really dry since the electrolyte is a moist paste. Dry cells are sealed to prevent the liquid in the electrolyte from evaporating. If the electrolyte does become dry, the cell becomes useless since the cell is no longer able to change chemical energy to electrical energy.

The two electrodes of the dry cell are carbon and zinc, with the carbon rod serving as the positive and the zinc as the negative electrode. The paste is a miniature chemical factory but its actual composition will vary, depending on the manufacturer. The paste mixture can contain ammonium chloride, powdered coke, ground carbon, manganese dioxide, zinc chloride, graphite, with enough water added to form a paste.

Dry cells are available in a large variety of sizes and current-delivering capabilities. As a general rule, the smaller the size of the cell, the smaller its possible current output, although there are some exceptions. One commonly used size

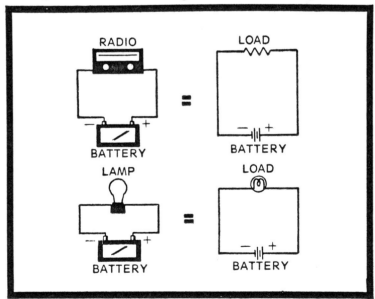

Fig. 2-9. Any device that draws current from a voltage source, such as a battery, is called a load.

41

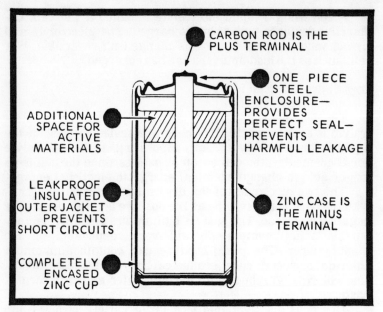

Fig. 2-10. The size D cell uses a center electrode of carbon and a container of zinc. The carbon rod is the plus electrode; the zinc, the minus electrode.

is the No. 6, about 2½ inches in diameter and about 6 inches long. When the cell is fresh, its output voltage, measured when connected and delivering current to the load, is about 1.5 volts. As the cell becomes older, due to chemical action in the cell, the terminal voltage (the voltage measured across the electrodes) will decrease, dropping to about 0.75 volt. This is the voltage measured under open-circuit conditions with the cell **not** delivering current to a load. With a load connected, the terminal voltage will be even less than 0.75 volt and can be as low as 0.5 volt. When the terminal voltage reaches this level, the cell is ready to be discarded.

Cell life can be increased by giving it some recuperation time between current demands. A cell that must deliver current constantly will not last as long as one that is allowed to work intermittently. The life of a cell not only depends on whether it is used on a continuous or intermittent basis, but also on the amount of current drawn from the cell. The No. 6 dry cell can supply a fair fraction of an ampere on a continuous basis for 10 hours or so, but it can also tolerate short-term drains of several amperes for circuits that go on and off. The short circuit current of a No. 6 dry cell is about 25 amperes. As the cell is used and as it gets older, its short-circuit current capability will decrease, dropping to 10 amperes or less.

The size D cell shown in Fig. 2-10 has essentially the same internal construction as the No. 6, but there are a few physical differences. This cell is sometimes known as a flashlight cell, but you will also find it used in other applications. The short-circuit current is about 6 amperes. A disadvantage of this cell is that it tends to corrode and swell after it is discharged and for this reason can damage equipment. Better-quality D cells are built with a steel jacket around the zinc container electrode. Such cells are more expensive, although they may not be able to deliver more voltage or current than an ordinary D cell.

Polarization

Dry cells can be afflicted by a condition known as polarization, which reduces the output voltage and current. When a dry cell is connected to a load and current is flowing, hydrogen bubbles collect around the carbon rod (positive electrode). These bubbles tend to insulate the carbon from contact with the paste electrolyte. To remove the hydrogen bubbles, oxygen in the form of manganese dioxide is mixed in with the electrolyte. Manganese dioxide has a rich supply of oxygen and releases enough of it so that it combines with the hydrogen bubbles to form water. Because of its action, the manganese dioxide is referred to as a depolarizing agent.

Nickel-Cadmium Cells

Not all small cells are carbon-zinc construction. Nickel-cadmium cells are available in the same size as small dry

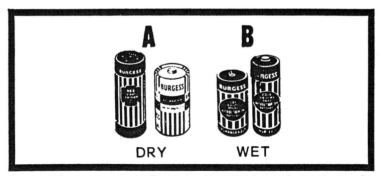

Fig. 2-11. Dry and wet cells. The wet cells are nickel-cadmium types; dry type are zinc-carbon. Nickel-cadmium cells can be charged; dry cells cannot.

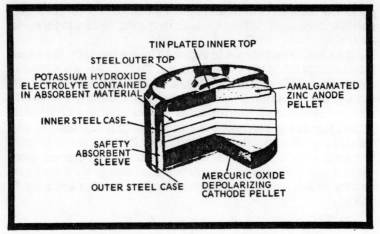

Fig. 2-12. Construction details of the mercury cell.

cells, (Fig. 2-11) but are true storage units (secondary batteries). They can be recharged many times. The type numbers carry the same letter designations as their carbon-zinc counterparts and are interchangeable with them. Thus, you can get nickel-cadmium cells in sizes AA, C, D, etc. The output of a single cell is 1.25 volts. Nickel-cadmium cells are also available in "button" size: only 29/64" x 7/32".

Larger size nickel-cadmium batteries are used in aircraft storage batteries, much as a lead-acid battery is used in an automobile. For a given size, they can store and deliver more electrical energy than the lead-acid type, and are also much lighter. With equal power ratings, for example, a nickel-cadmium unit is about one-third lighter than the lead-acid cell.

Silver-Oxide Button-Type Cells

These are used in hearing aids and electric watches. They supply 1.5 volts and are approximately 0.455" x 0.165", although other sizes are available.

The Mercury Cell

There are a large number of different types of dry cells, all intended for a variety of applications. One of these is the mercury cell, (Fig. 2-12) engineered to produce a cell of small size which would last a long time, but which would also be highly dependable. The cell, housed in a small steel case, uses an electrolyte of potassium hydroxide contained in an absorbent material. In this respect, the unit resembles the

Edison battery. The positive plate is composed of amalgamated zinc while the negative consists of pellets of mercuric oxide, from which the cell derives its name. Although the word plate is in common use, possibly due to the fact that both the lead-acid battery and the Edison battery use plates, the electrodes can have other shapes: they may be cylindrical as in the case of the No. D dry cell, or in granular or pellet form as in the instance of the mercury cell. The voltage of the mercury cell is about 1.34 volts and it maintains this voltage throughout most of its useful life.

Mercury cells are often connected to form a battery. The cells are packaged in various kinds of cases to supply a number of different voltages: 5.4 volts; 6.75; 8.4; 9.8; 12.6, and so on.

Wet Cells

Wet cells offer a huge advantage over dry cells in general because they possess the capability of being charged time after time. But the charging process can create problems under some conditions, because wet cells tend to "gas" during charging—particularly when a charge is continued after the battery has been restored to its original condition.

This gassing is the formation of hydrogen at one of the electrodes; it escapes through the electroyte and is vented, normally, through the port where water is added.

For a single battery, the amount of hydrogen released is rather small, but in commercial installations where large numbers of batteries are charged at the same time, the amount of hydrogen is sufficient to constitute a danger. Hydrogen is explosive, and so there are usually printed warnings not to smoke or light matches in the immediate region of the charging batteries. As an added safety procedure, the batteries may vent their hydrogen into a metallic hood to permit the hydrogen to escape into the outer air.

Lead-Acid. Of the various wet cells used today, the lead-acid type is probably the most common. A battery of such cells is used in automobiles to supply starting power. Invented by Gaston Plante in 1859, it consists of two lead plates immersed in a watered-down solution of sulphuric acid. One of the plates is made of pure sponge lead and is the negative electrode. The other is lead peroxide and is the positive electrode. The potential developed by this cell is approximately 2.1 volts.

During the time a load is connected across the terminals of the cell and current flows out of it, a chemical reaction takes place during which the pure lead plate changes to lead sulphate. The electrolyte ionizes (changes into ions of hydrogen and sulphate). The sulphate ions migrate to the lead plate while the hydrogen moves toward the lead peroxide plate. In the chemical process, the pure lead plate accumulates electrons while the lead peroxide plate loses them. When a load is connected to the cell, a movement of electrons takes place from the negative electrode, through the connecting wires and load, back to the positive electrode.

The lead-acid cell can be recharged. The charging source supplies a voltage that is slightly higher than that of the lead-acid cell, forcing current through the cell in a direction opposite the flow of current from the cell through the load. During the discharge process, the electrolyte becomes less and less acidic, gradually changing to water. Both electrodes also change from lead and lead peroxide to lead sulphate. However, during the time the cell is being charged, the electrolyte becomes more and more acidic and the electrodes go back to their original form of lead and lead peroxide.

The Edison Cell. Another type of cell, but not as widely used as the lead-acid type, is the cell invented by Thomas Edison and sometimes known as the nickel-iron-alkali cell. The positive electrode contains nickel hydroxide while the negative electrode uses iron oxide in finely divided form, mixed with some cadmium. The electrolyte in this cell is a base material rather than acid and consists of a 21 percent solution of potassium hydroxide, in water, to which a small amount of lithium hydroxide is added. The output potential of an Edison cell, when fully charged, is about 1.3 volts.

The Edison cell is somewhat lighter and stronger than the lead-acid type, can be recharged more rapidly, has a longer life, and isn't damaged by overcharging or by complete discharge. For the most part, it is used in industrial applications.

Both the lead-acid cell and the Edison cell are wet cells, are secondary types, and can be recharged.

VOLTAGE OF A CELL

The voltage, or emf, of a cell is determined by the materials used for the electrodes. The size of the electrodes has no effect on the voltage, and so a cell that is rated at about 2 volts may be so small that a half-dozen of them could fit into the palm of your hand. Alternatively, a 2-volt cell could be ten times the size of this book. The average voltage of a cell, regardless of size or the type of electrodes used, ranges from about 1 volt to 3 volts.

Increasing the Voltage

Since cell voltage is so small, it is sometimes necessary to connect cells to supply a large emf. The technique used is known as a series connection (Fig. 2-13) and the resulting cell combination is a battery. In this type of connection, a conductor is connected from the minus terminal of the first cell to the plus terminal of the next. Generally, only identical cells are wired in series. The total voltage is equal to the voltage of any of the individual cells multiplied by the number of cells. Thus, if five 1.5-volt cells are wired in series, the total available voltage is equal to 5 x 1.5, or 7.5 volts. Cells in series are often supplied with the series connection made by the manufacturer. The battery will then consist of cells placed in some kind of container and may also have some kind of sealing compound to keep the cells in place. In this arrangement only two external terminals will be visible: a positive and a

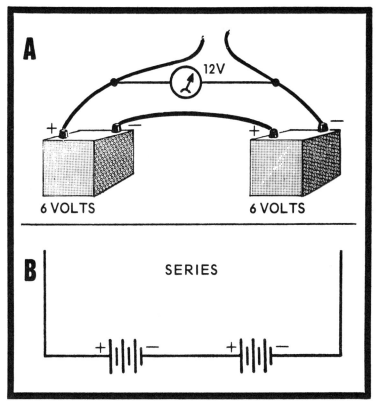

Fig. 2-13. When cells are connected in series we get more voltage. The current capacity is not increased. Pictorial (A) and circuit (B).

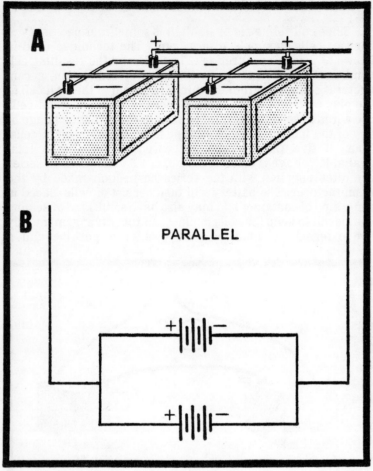

Fig. 2-14. When cells are connected in parallel, current capacity is increased; voltage remains unchanged. Cells should be identical, with the same amount of charge. Pictorial (A) and circuit (B).

negative terminal, even though the individual cells inside the container have their own terminals. Thus, it is possible to purchase 9-volt batteries, 15-volt batteries, and so on. When cells are wired in series, only the total voltage is increased. The maximum available current is that of a single cell in the series group.

CURRENT OF A CELL

The amount of current delivered by a cell depends on the amount of material in the electrodes and the surface area exposed to the electrolyte. And so a cell that is required to be able to furnish a relatively large amount of current is larger and much heavier than one that can supply only a small amount of current.

To increase the amount of current that can be obtained from cells, they can be wired in parallel, as shown in Fig. 2-14. In this type of connection, all the positive terminals of the cells are joined, and all the negative terminals are wired together. (Positive and plus are synonymous terms, as are negative and minus.) It is always best to connect identical cells in parallel. When cells are connected in parallel, the total available voltage is that of any one of the cells, but the total available current is the current capability of a single cell multiplied by the number of cells in the parallel circuit.

The Battery's Load

All electrical components that are wired to batteries, such as lamps, or radio receivers, or motors, have definite voltage and current requirements to enable them to work properly. Depending on the way it is designed and manufactured, a device may require a small amount of voltage but a large current, or a large voltage and a small current. It is also possible for a load, such as a small lamp, to need a small amount of voltage and also a relatively small amount of current. Quite often the device will be marked with its voltage requirement, such as 2V, 5V, etc. Normally, the current demand of the device isn't indicated.

And so, the kind of battery you can use is dictated by the kind of load you will connect to the battery. If the battery is unable to supply the right amount of voltage, or current, or both, the device will either not work at all, or may work improperly.

To obtain a higher voltage and a greater current than can be supplied by a single cell, it is customary to connect them in a series-parallel arrangement, as shown in Fig. 2-15. This is simply a combination of the series and parallel circuits described earlier. The number of cells connected in this way is determined by the voltage and current demands of the load. In an automobile, for example, the starter motor requires about 12 volts, but an extremely heavy current. The car battery contains 6 cells wired in series, with each cell having a maximum output voltage of about 2.1 volts. The 6 cells in series supply 6 x 2.1, or 12.6 volts. To be able to supply the large amount of current required, the cell electrodes are quite large and also have a large surface area exposed to the acid electrlyte.

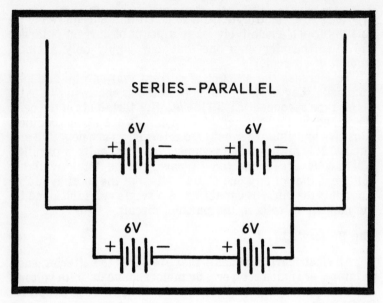

Fig. 2-15. We can get more voltage and more current by using a series-parallel circuit. Here the total output is 12V; the total current is that of two of the batteries.

THE UNIT OF CURRENT

With every cell we are concerned with two important factors—voltage and current. Voltage is electrical pressure or emf and its basic unit is the volt.

Since the electron is so extremely small, tremendous amounts of them must flow to supply a measurable current. A coulomb of electricity is an easily measured amount, for it consists of 6.28×10^{18} electrons, or 6,280,000,000,000,000,000 electrons. This number of electrons moving past a given point in one second constitutes a current of one ampere. The term "ampere" is used more often in electricity and electronics since we are invariably concerned with moving currents. The ampere is the basic unit of current. It is often abbreviated as A, amp, or amps. 6A means 6 amperes.

INTERNAL RESISTANCE OF A CELL

As a cell is used the electrolyte changes chemically. One effect of this chemical change is that the electrolyte becomes less conductive. Thus, in the lead-acid cell the electrolyte gradually changes from acid, a good conductor, to water, a much poorer conductor.

Fig. 2-16 shows a load connected across a cell. External to the cell, the current flows from the negative terminal, through the connected load, to the positive terminal of a cell. Inside the cell, the current flows from the positive electrode of the cell, through the electrolyte, to the negative electrode. The electrolyte acts as a conductor inside the cell, just as the connecting wires act as conductors outside it. But as the cell becomes used the electrolyte begins to oppose the flow of current through it. This opposition is known as the internal resistance of a cell. If the cell is completely discharged, this opposition to the passage of current may be so great that no current will be supplied to an external load. A practical example is the inability of a dead battery to start a car. The

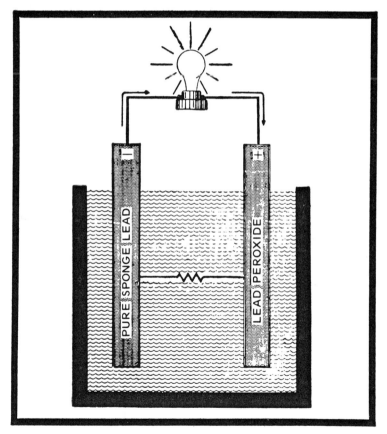

Fig. 2-16. As a cell becomes used, the electrolyte becomes less conductive—that is, the internal opposition or resistance to the flow of current increases. The internal resistance is represented here by the load symbol.

battery may still have a substantial amount of voltage, but it has little or no current-delivering capability.

Recharging a secondary cell has the effect of lowering the internal resistance of the cell because the electrolyte becomes more conductive during the charging process. In the lead-acid cell, the electrolyte becomes more acidic.

BATTERY DUTY

In the case of carbon-zinc cells and nickel-cadmium types, the smallest size has the same voltage as the largest. The difference between size C and D is not in voltage, but in current delivering ability. The larger the cell, the greater the size of its electrodes. A cell that can deliver more current for a longer time than a comparable cell, but smaller in size, is called a "heavy duty" unit. A heavy-duty automobile battery has the same terminal voltage as other auto batteries, but is capable of supplying a larger amount of current and will be able to do so for a longer time. Naturally, such batteries cost more than ordinary types.

LIFE OF A BATTERY

Batteries are made to be used. All that storing a battery will do will be to delay somewhat the inevitable end. If a dry cell isn't put to work, it will deteriorate, consume itself, and will have no further value. In the case of rechargeable types of cells, the cell can be rejuvenated by recharging, but even here, storage reduces the life of the cell.

A cell or battery that isn't used will gradually wear out because of chemical changes. The time during which a cell or battery can be stored, but still have useful life (that is, the cell or battery can be put to work) is known as "shelf life." As a rule, the smaller the size of the cell, the shorter its shelf life. Alkaline cells can have a longer shelf life than carbon zinc types. (Shelf life is increased if the storage area is kept cool.)

The length of life of a dry cell depends on its construction and the way in which it is used. A cell that is used intermittently will last longer than one that has a constant current drain. The maximum current available from a cell is known as its short-circuit current.

Short-circuit current simply indicates the greatest current drain possible and does not mean "working" current, a current that is far below short-circuit current. Short-circuiting a battery can ruin the unit since the large resulting current can overheat the cells. If the battery is a large size, such as an automobile lead-acid type, a short circuit can be dangerous. It has been estimated that the energy in a heavy-duty lead-acid cell is capable of lifting a ton one mile into the sky.

Local Action

It may be surprising that a cell will be "at work" even when it is not connected to a load and not delivering current. In theory, when a switch is opened and the battery is detached from its load, there should be no further current flow. Actually, however, current does continue to flow, but only inside the cell. In the case of the dry cell, various impurities may exist in the zinc container, also acting as the negative electrode. The zinc that is used isn't pure, for obtaining pure zinc is a difficult and expensive process, but contains impurity traces of iron, carbon, lead and arsenic. Since, by definition, a cell consists of two dissimilar materials in the presence of an electrolyte, these impurities and the zinc represent tiny cells. The current flow is known as local action, and while the amount of current circulating in the zinc electrode is small, it is continuous. Local action will ultimately result in the zinc being completely consumed, even though the cell may never have been taken off the shelf.

Local action takes place in wet cells also, but since the wet cell can be recharged, the effects of local action can be overcome.

MAKING BATTERY CONNECTIONS

The larger the amount of current delivered by a battery, the more important do connections become. In an auto battery, for example, the connections must not only be tight, but must cover as much of the external battery terminal as possible. In addition, for maximum current flow, there must be as little corrosion as possible between the battery terminals and the connecting cables. The cables, of course, must be large enough to carry the current demanded by the load from the battery.

As a battery becomes smaller and delivers less current, connections can be of the pressure type, as in the case of flashlights or transistor radios. The cells are usually connected automatically by inserting them in specially designed holders. Not only do the holders supply the connection to the external circuit, but are also designed to hold the cells so they are properly joined in either a series or parallel arrangement.

MORE ABOUT THE ELECTRON

The electron is the sum and substance of the sciences of electricity and electronics. Thus, it is important to learn more about producing and controlling it. Chapter 3 is a step in that direction.

How the Electron Was Finally Trapped

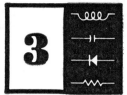

Electricity was put to practical use long before investigators knew what it was. Electrical currents were controlled because some facts had been learned about current behavior. Benjamin Franklin regarded electricity as consisting of extremely small particles (right!) moving from positive to negative (wrong!). However, Franklin's insight into the nature of electricity was remarkable. Even his wrong-way theory about electricity was practical—for many years. Franklin believed there were two kinds of electricity since he was able to make certain objects repel or attract each other, when rubbed. Again, an incorrect supposition but quite a logical deduction based on his experiments (Fig. 3-1). This concept of two kinds of electricity still persists today as shown in any discussion of positive and negative charges. Actually, the terms positive and negative are relative, depending entirely on electron quantity. Imagine two plates (Fig. 3-2) each holding different amounts of electrons. The plate with the fewer electrons is called positive (or is said to be positively charged) with respect to the other plate. But it still has electrons. The plate having the larger number of electrons is negative—not negative by itself, but with respect to the other plate. Again, a matter of electron quantity. Any charged plate can be both positive and negative, depending on its reference.

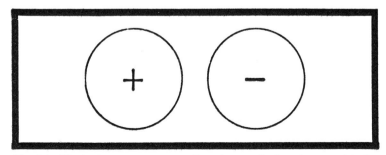

Fig. 3-1. All electrons carry a negative charge. A plus sign, used to indicate a positive charge, simply means a substance that has fewer electrons than some other substance.

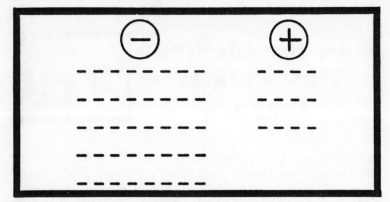

Fig. 3-2. A plate having a larger number of electrons than some other plate is called negative. The plate with the smaller number of electrons is called positive. The words positive and negative are relative.

There is nothing wrong with talking about positive charges, provided it is understood that this simply means fewer electrons in comparison with some object that has more electrons.

THE NATURE OF MATTER

To understand electricity, and its descendent, electronics, it is necessary to probe into the nature of matter. All matter is made of elements. Most elements are natural, but some are man-made, created in scientific laboratories. Quite a few elements—gold, silver, aluminum, iron, tin, hydrogen, helium, neon, may be familiar. An element can be a solid such as iron, a liquid such as mercury, or a gas such as oxygen or chlorine. An element is a substance having identical atoms. Aluminum is an element because all of its atoms are the same. Aluminum can be subdivided until each part is practically invisible, yet each of the parts is still aluminum. The only way in which aluminum can be changed into some other form is to combine it with another element. Every element has certain properties which identify the element—properties such as inertia, volume, weight, mass, density.

The simplest of all the elements is hydrogen. One concept of this gas is shown in Fig. 3-3 and consists of a central portion called the nucleus and one electron which revolves in an orbit around the nucleus. This orbit is sometimes referred to as a shell, or a ring, and is often identified by the letter K. Hydrogen has a K shell comprising a single electron. The K shell is also sometimes called the first energy shell or energy level.

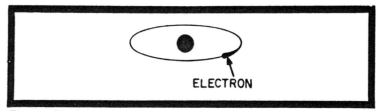

Fig. 3-3. The hydrogen atom consists of a nucleus and a single electron revolving in orbit.

That electron traveling around the nucleus is the part of the atom we are interested in. No one has ever seen an individual electron but if you've ever seen an electric spark or a flash of lightning, you may still not have seen electrons, but the results of what they can do.

The electron has mass; it can be weighed. And because it has mass it can be called a particle. But this particle is most unusual. It carries a negative charge. Further, all electrons are alike, and since they all carry a negative charge, repel each other. Electrons do not like to share with other electrons and so if electrons are forced onto a metal plate, will flow from that plate along a wire to some less electron-packed plate (Fig. 3-4). It is this movement of electrons we call an electric current.

A substance will move if pressure is applied to it. It is the force of repulsion among electrons that compels them to move and which we refer to as voltage. A battery supplies voltage because one of its plates or electrodes is crowded with electrons while the other electrode has far fewer electrons. This difference in electron quantity is voltage. Voltage is nothing mysterious and is just the pressure exerted by electrons, caused by their repelling effect.

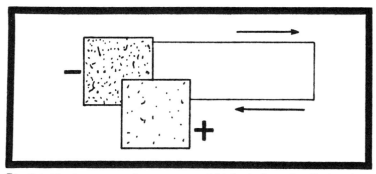

Fig. 3-4. Electrons move from an area of high electron density to a less electron-crowded region. The arrows indicate the direction of current flow through the connecting wire.

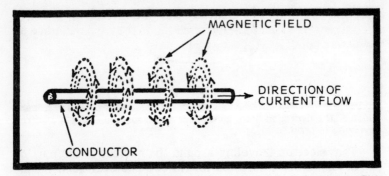

Fig. 3-5. A magnetic field surrounds a current-carrying wire. The magnetic field is at right angles to the conductor, with the strength of the field dependent on the strength of the current.

There is still another important property of electrons. Put them in motion and they each become tiny magnets (Fig. 3-5). The magnetic fields around the individual electrons cooperate, with the total magnetic strength dependent upon the amount of current flow. A weak current—that is, a current consisting of a relatively small number of electrons—will result in a very weak magnetic field, relatively few lines of magnetic force surrounding the wire carrying the current. Increase the current, force more electrons to flow through a

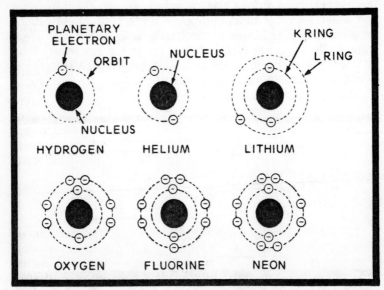

Fig. 3-6. Electrons travel in rings or orbits around the nucleus of the atom. Each ring can have a certain maximum number of electrons.

wire, and the magnetic field around that wire will become much stronger. The magnetic field around the wire will have a north and a south pole, just like a magnet.

OTHER ELEMENTS

Elements can be arranged in order, depending on the number of electrons they have in orbit. Helium (see Fig. 3-6) has two orbital electrons in its K shell. The K shell, though, has an electron limit and will not accommodate more than two electrons. Element lithium has three electrons but the third electron is in a separate orbit of its own. This second orbit is known as the L ring. Unlike the K ring, the L ring may contain as many as 8 electrons. Following lithium we have beryllium with two electrons in the K ring; carbon has four, nitrogen five, oxygen six, fluorine seven, and neon has eight. All of these elements, of course, have two electrons in the ring closest to the nucleus.

After a ring has reached a certain number of electrons, a new ring is formed. The K ring has a maximum of two and the L ring a maximum of eight. The third shell is called M and can have a maximum of 18 electrons, and the N ring, surrounding the M, a total of 32.

COMPOUNDS

An electron ring that does not have the maximum number of permitted electrons is an incomplete ring. A pair of elements can combine their outermost electrons but in so doing they form a new substance called a compound. Pure iron is a white metal but when united with oxygen it is more familiarly known as rust. In another example, the outer electrons of oxygen and the single orbital electron of hydrogen will combine to form water. Oxygen and hydrogen are gases; water is a liquid. Oxygen and hydrogen support combustion; water is used to douse fires. Combinations of elements can become somewhat complicated.

MOLECULES

The union of atoms results in the formation of a substance called a molecule. The molecule can consist of identical atoms or atoms of different elements. The molecule can have properties which are quite different from those of the atoms from which it is formed. Fig. 3-7 shows the development of a molecule of salt from sodium and chlorine atoms.

IONIZATION

An atom can share its outermost orbital electrons with another atom, forming a compound, or it can gain or lose some of those orbital electrons. An atom that gains or loses electrons is ionized—that is, it is no longer neutral as far as charge is concerned. If an atom gains an electron it is more negative than before the acquisition and so is no longer called an atom but an ion. And since it is more negative, due to the extra electron, it is known as a negative ion. It is negative because electrons carry a negative charge.

An atom can also lose one of its orbital electrons. In losing an electron (negative charge) the atom is now called an ion. And, since it has lost a negative charge it is positive.

The process of changing atoms into ions is called ionization. A substance that is normally a good insulator can be changed into a good conductor by ionization. Pure water is a poor conductor. When salt is put into the water, the salt dissolves and some of its atoms ionize. Salt is composed of sodium and chlorine. Salt, the compound, is sodium chloride.

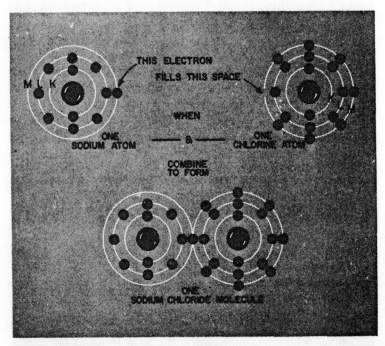

Fig. 3-7. An atom of sodium (a metal) and an atom of chlorine (a posionous gas) combine to form a molecule of sodium chloride (salt) by sharing outer orbital electrons.

Immersed in water, some of it separates into sodium and chlorine ions—no longer atoms of sodium and chlorine, but atoms which have gained and lost electrons respectively. The ionization of the water changes it from a poor conductor to a good one. In the lead-acid cell, the electrolyte is sulphuric acid in water. When the acid is put into the water, the compound partially breaks down into hydrogen ions and sulphate ions, and so the water solution of suphuric acid becomes ionized. Ionization is a technique sometimes used in gas-filled tubes as a means of supplying a low-resistance path for the current flowing through the tube.

BOUND AND FREE ELECTRONS

A force of attraction exists between the nucleus of an atom and the electrons revolving in orbit around it. The force of attraction, however, is weakest for those electrons in the orbit furthest removed from the nucleus, hence it is possible to strip an electron from the outermost ring. However, electrons can also exist independently of atoms. Wipe a dry glass window with a cloth and you transfer of billions of electrons from the cloth to the glass. Electrons not directly bound to atoms are called free electrons.

ELECTRON STATISTICS

An electron is so small and so light that the lightest of all atoms, the hydrogen atom, weighs about 2,000 times as much as an electron. The electron has been estimated to have a diameter of 1×10^{13} inch, or 0.0000000000001 inch. It would take 1,000,000,000,000 electrons to form a line only 1 inch long. We can make electrons remain at rest or we can accelerate them until they reach a speed of about 150,000 miles a second.

All electrons are identical and all carry a negative charge. The charge on the electron is negative and exists whether the electron is in motion or not. Each electron is also a tiny magnet and if we get enough of them moving through a conductor we can detect the magnetic field produced by the electrons—and further, are able to put that magnetic field to work.

Electrons are matter and have the properties of matter. Thus, electrons have inertia and so will remain at rest or in uniform motion in the same straight line or direction unless acted upon by some external force. This means electrons oppose or resist being set in motion but once in motion resist being stopped. You've had personal experience with inertia if you've ever tried to push a stalled automobile. An auto at rest wants to stay that way, but once the wheels start turning not so much push is required to keep the car in motion, assuming a flat, smooth stretch of road. If you lurch during a sudden stop while holding on to a strap in a crowded bus, subway or train, that's inertia again. Your body, in motion, tends to keep moving when the brakes of the moving vehicle are applied.

AMPERES, MILLIAMPERES, MICROAMPERES

For the most part, electricity and electronics are concerned not with electrons at rest but with electrons in motion. When one coulomb (6.27×10^{18} electrons) flows past a given point in one second, this moving mass of electrons is referred to as an ampere. The ampere is the basic unit of current flow. Formulas in electricity and electronics that involve current flow use the ampere as the basic unit. The difference between the coulomb and ampere is one of movement or nonmovement for they both involve the same quantity of electrons. The coulomb specifies a quantity of electrons but has no reference to motion. An ampere involves the same quantity of electrons, but now the electrons are traveling through some conductor, such as a wire, or may be moving through a gas, such as mercury vapor, or may be traveling through a vacuum, such as the vacuum in a tube.

Although the ampere does represent a large number of electrons in motion, it isn't an uncommon unit. An electric heater or electric toaster may have currents of several amperes. The starting current of an automobile can easily exceed a hundred amperes. The current for a motor is several amperes, usually more. However, while large currents aren't unusual, quite often an electrical or electronic device may required currents of less than one ampere.

The first submultiple of an ampere is the milliampere (abbreviated as mA) or thousandth of an ampere (Fig. 3-8). If an ampere could be divided into a thousand equal parts, each of these equal parts would be one milliampere. In arithmetical terms:

1 ampere = 1,000 milliamperes = 1,000 mA

Electrical and electronics technicians often find it necessary or convenient to convert from milliamperes to amperes or from amperes to milliamperes. The conversion process isn't difficult but it does require some arithmetic.

Example:

A current of 635 milliamperes flows through a wire connecting two components. What is the equivalent current in amperes? 635 milliamperes is the same as 635. milliamperes. To convert to amperes, move the decimal point three places to the left, equivalent to dividing by 1,000.

$$635. \text{ milliamperes} = 0.635 \text{ ampere}$$

Example:

A current of 25 milliamperes flows through a tube used in a television receiver. What is the equivalent current in amperes? 25 milliamperes can also be written as 25. milliamperes. To convert to amperes, divide by 1,000 or move the decimal point three places to the left.

$$25 \text{ milliamperes} = 25/1000 \text{ ampere}$$
or
$$25. \text{ milliamperes} = 0.025 \text{ ampere}$$

The conversion of amperes to milliamperes involves multiplying amperes by 1,000 or moving the decimal point three places to the right.

1 ampere	1,000 milliamperes
1 ampere	1,000,000 microamperes
1 milliampere	1/1,000 ampere
1 milliampere	1,000 microamperes
1 microampere	1/1,000,000 ampere
1 microampere	1/1,000 milliampere

Fig. 3-8. The ampere and its submultiples.

Example:

A current of 3 amperes flows in a relay circuit. What is the equivalent current in milliamperes?

```
3 amperes = 3.0 amperes = 3,000 milliamperes
or
3 amperes = 3 x 1,000 = 3,000 milliamperes
```

Example:

A current of 1.25 amperes flows through the filament of a transmitting tube. What is the equivalent current in milliamperes?

```
1.25 amperes = 1.25 x 1,000 = 1,250 milliamperes
or
1.25 amperes = 1,250. milliamperes (obtained by
moving decimal point)
```

There are some electrical and electronic circuits that require currents that are less than one milliampere. While the milliampere as a current unit would be used to describe such currents, it would involve awkward fractions such as 3/17 milliampere or 28/93 milliampere. This could be simplified by using decimals, but this would also involve awkward numbers requiring zeros. We would have currents such as 0.00045 milliampere or 0.000065 milliampere. To be able to use more convenient whole numbers, another submultiple of the ampere is available and is known as the microampere. The microampere is abbreviated as uA. The lower case u represents the greek letter "mu," and is often used in electrical and electronic formulas. Thus, 10 microamperes equals 10 uA.

The microampere is equivalent to one one millionth of an ampere or one thousandth of a milliampere. Divide an ampere into a million equal currents and each would be one microampere. Or, divide a current of one milliampere into a thousand equal currents and each will be one microampere. Electrical and electronics technicians are often required to convert back and forth between amperes, milliamperes, and microamperes.

Current Conversion Rules

As a general rule, when moving from a larger quantity to a smaller quantity, multiply. And, when moving from a smaller quantity to a larger, divide.

To convert amperes to milliamperes, multiply amperes by 1,000. To convert amperes to microamperes, multiply amperes by 1,000,000. To convert milliamperes to microamperes, multiply milliamperes by 1,000.

Note in each of the above conversions, the move was from a larger quantity (the ampere) to a smaller quantity (milliampere or microampere).

To convert milliamperes to amperes, divide milliamperes by 1,000.

To convert microamperes to amperes, divide microamperes by 1,000,000.

To convert microamperes to milliamperes, divide microamperes by 1,000.

Example:

What is the equivalent current, in microamperes, of a current that is measured at 22 milliamperes?

22 milliamperes x 1,000 = 22,000 microamperes
= 22,000 µA.
22 milliamperes = 22. milliamperes.

Moving the decimal point three places to the right gives 22,000 milliamperes.

Example:

A current of 3,645 microamperes flows in a transistor circuit. What is the equivalent current in milliamperes? In amperes?

3,645 microamperes = 3,645. microamperes = 3,645 µA.

To convert to milliamperes move the decimal point three places to the left. Hence:

3,645. microamperes = 3.645 milliamperes = 3.645 mA.

To convert 3,645 microamperes to amperes, divide by one million or move the decimal point 6 places to the left.

3,645. microamperes = 3,645/1,000,000 ampere
or
3,645. microamperes = 0.003645 ampere = 0.003645A.

THE AMPERE-HOUR

The coulomb per second, or ampere, uses the second as its basic unit of time, but the second is quite a small time unit. However, in addition to the coulomb per second, we have the ampere-hour. Since there are 60 seconds in 1 minute and 60 minutes in 1 hour, 60 x 60 equals 3,600. Hence the ampere-hour is a unit that is equivalent to 3,600 coulombs. The ampere-hour (A-hr) is a term often used in connection with battery current capability or output. A 100 A-hr battery is not a battery that can deliver a current of 100 amperes for one hour, but it can deliver roughly that equivalent over a much longer period of time—say one ampere for 100 hours. As the current goes up, the capability of the battery goes down. So the same battery could not supply 50 x 2 equals 100 A-hr—that is, it could not deliver 2 amperes for 50 hours, though it could approach this figure.

Excessive heat generated inside the battery and efficiency losses because of these higher temperatures prevent large current outputs for extended periods. Actually, the ampere-hour rating of a battery is meant to specify the requirement for bringing a spent battery to a fully charged state. But at low current levels (without the presence of heat) the discharge and charge current-and-time values are similar.

THE MEANING OF VOLTAGE

Voltage and current are often considered as two separate quantities or units, and it is convenient and practical to do so. Voltage is electrical pressure and is often compared to water pressure. Water held in by a retaining wall, such as a dam, exerts a pressure against it. When the gates to the dam are opened, the water goes from a static (at rest) condition, to a dynamic (moving) condition. However, the water hasn't changed. The pressure and the movement are both conditions of the same substance.

Electrons also exert pressure since all electrons carry a negative charge. Because similar electric charges repel each other, a collector of electrons exerts a potential pressure, more technically known as voltage, or electromotive force (emf). In a battery the difference in electron quantity on two electrodes is an electrical pressure or voltage. Voltage always exists between any two areas that have different quantities of electrons. A battery is just one example. A glass rod, rubbed with a cloth, is charged because the electron quantity on its surface is different from that of a nearby substance.

Whenever there is a difference in electron quantity, the electrons will attempt to equalize matters by distributing themselves uniformly on all surfaces. In the case of a battery, the negative electrode has an electron surplus; the positive electrode simply has fewer electrons. Connect a wire between the two electrodes and a current will flow from the negative to the positive electrode.

Voltage Units

The basic unit of electromotive force or electrical pressure is the volt. There is a relationship between voltage and current, but a large voltage doesn't necessarily mean a large current. A dam might have a pressure of thousands of tons against it, but the water control can be such that just a small trickle flows.

The ampere is the basic unit of current and so the milliampere and the microampere are submultiples. In the case of voltage, however, there are commonly used units that are larger and smaller (Fig. 3-9).

1 volt	1,000 millivolts
1 volt	1,000,000 microvolts
1 volt	0.001 kilovolt
1 volt	0.000001 megavolt
1 kilovolt	1,000 volts
1 kilovolt	0.001 megavolt
1 megavolt	1,000,000 volts
1 megavolt	1,000 kilovolts
1 millivolt	0.001 volt
1 millivolt	1,000 microvolts
1 microvolt	0.000001 volt
1 microvolt	0.001 millivolt

Fig. 3-9. Voltage multiple and submultiple conversions.

The kilovolt is equal to 1,000 volts while the megavolt is equivalent to 1,000,000 volts. Moving in the other direction, the millivolt is the same as one-thousandth of a volt while the microvolt is one-millionth of a volt. It is often necessary for electrical and electronics technicians to be able to move easily and quickly between these voltage multiples and submultiples. The technique for doing so, however, is exactly the same as that described earlier for current submultiples.

Voltage Conversions

As in the case of current, conversion from a larger to a smaller quantity involves multiplication; while conversion from a smaller to a larger quantity requires division.

To convert megavolts to kilovolts, multiply megavolts by 1,000.

To convert megavolts to volts, multiply by 1,000,000.

To convert kilovolts to volts, multiply by 1,000.

Each of these rules involves conversion from a larger to a smaller unit and so multiplication is involved.

To convert from volts to kilovolts, divide volts by 1,000 and to convert from volts to megavolts, divide by 1,000,000.

To convert from volts to millivolts, multiply volts by 1,000. The conversion of millivolts to volts requires division by 1,000. Finally, the conversion of volts to microvolts is done by multiplying by 1,000,000 while the change from microvolts to volts means dividing by 1,000,000.

ABBREVIATIONS

The letter V is logically used as an abbreviation for voltage and so a designation of 200 volts can be written as 200V. Another letter, E, is also used to represent voltage although it is actually an abbreviation for electromotive force. The E is used to mean the force itself. While the V is used with a number to denote a specific quantity. Thus, we could say, for example, E equals 100V.

Recall that A is an abbreviation for amperes. 5A is 5 amperes. I is used to denote current in general, though. So we could say, I equals 5A.

Sometimes a letter is used as a multiplier. The lowercase letter k is a notable example and indicates multiplication by 1,000. 1,000 volts can be written as 1 kV. The k is an abbreviation of the prefix kilo.

You will note that many letters in electronic units of measure are capitalized, while others are not. The electronics industry today capitalizes letters that honor men. Thus, mA, dB, mW, kV, and mH are used as abbreviations for milliampere (Ampere), decibel (Bell), milliwatt (Watt), kilovolt (Volta), and millihenry (Henry).

Sometimes, though, letters are capitalized for simple clarity. The lowercase m is already used to designate "thousandth," so a capital letter is used for "mega," or million. Thus, MV stands for megavolt, while mV represents millivolt.

A lowercase "u" symbolizes the Greek letter mu, and is an abbreviation of the prefix micro; it means one millionth. A microampere, written as uA, is a millionth of an ampere. One uV is one millionth of a volt.

CONDUCTORS

We now have two very important factors you will always find in electronics: voltage and current.

Electric currents are able to pass easily through some substances and only with great difficulty through others. A substance or material that allows the comparatively easy passage of electrons through it is called a conductor. The word comparative is used since no two substances are equally conductive. Silver is an excellent conductor. Copper is almost as good. Aluminum and zinc are next in order. Some liquids and gases are also conductive. Mixtures of liquids have various amounts of conducting ability, such as a salt water or a mixture of acid and water.

Offhand, it might seem desirable to have maximum conductivity and, under certain circumstances, this is correct. But just as it is desirable to control current flow, so too is current opposition or electrical friction sometimes deliberately introduced.

INSULATORS

An insulator is a substance which has a high opposition to the passage of an electrical current through it. Materials such as glass, porcelain, plastic, rubber, dry wood, all have insulating properties—that is, they can be depended upon, more or less, to oppose the flow of current through them. Pure water is an almost perfect insulator. Add some ordinary table salt and the water becomes a conductor.

While it is convenient to group substances as conductors or as insulators, all materials have a certain amount of conductivity; it is just a matter of degree. The conductivity of a

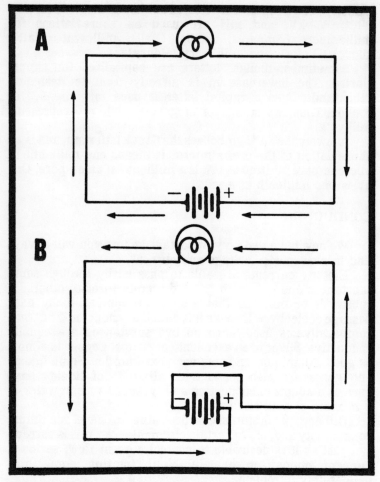

Fig. 3-10. The direction of current flow can be changed by transposing the leads to the battery, but the movement of electrons remains from minus to plus.

substance also depends on the amount of voltage across it. If the voltage is sufficiently high even a substance that is normally an insulator will conduct.

THE DIRECTION OF CURRENT

Current, like automobile traffic, can be made to flow in any direction, if properly guided. In the circuit of Fig. 3-10 the current shown in drawing A flows from left to right through the lamp connected to the battery. The direction of current is always from minus to plus in any load connected to a voltage source, such as a battery. Drawing B in the same illustration shows that the direction of current can easily be reversed by transposing the connections to the battery. In drawing A the current flows in one direction only and does not change its direction until the connections to the voltage source are changed. Similarly, the current in drawing B will also flow in one direction only, unless the connections to the battery are changed. In either instance, the current is called a direct current, abbreviated dc. The voltage that produces a direct current is also called dc, and so the abbreviation can refer to either voltage or current. Similarly, a nonalternating voltage, such as a battery, is called a dc voltage.

The direction of current is often indicated by an arrow with the head of the arrow pointing in the direction of flow of the current. The head of the arrow is sometimes marked with a plus (+) symbol and the opposite end of the arrow with a minus (−) symbol to emphasize that the current is moving from minus to plus.

ALTERNATING VOLTAGE AND ALTERNATING CURRENT

The voltage and current produced by power stations (Fig. 3-11) is alternating, a word that simply means the voltage reverses its polarity regularly and so the current keeps changing its direction. Alternating voltages and currents are described by the abbreviation ac. These voltages and currents can be produced by electromechanical generators or by devices which are completely electronic. Fig. 3-12 is a diagram of the way an ac generator works. For half the time the polarity of the generator is as indicated in drawing A. For the other half of the time the polarity reverses, as in drawing B. Note how similar this action is to the battery and its transposed leads shown earlier in Fig. 3-10. The difference is that the transposition is done automatically in the ac generator whereas with the battery it was necessary to physically connect and disconnect the leads.

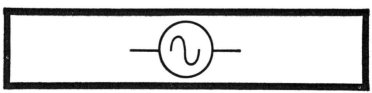

Fig. 3-11. Electrical or electronic symbol for an ac generator, which can be an electromechanical type or one that is purely electronic.

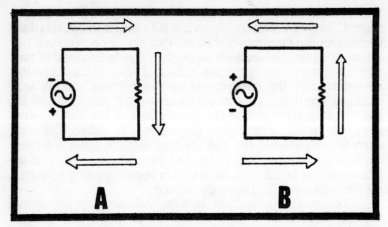

Fig. 3-12. Since the polarity of the voltage of an ac generator keeps reversing itself, the flow of the current through the conductors and the load is first in one direction, then the other.

When the polarity of the ac generator changes (Fig. 3-13), the direction of current reverses. The reason for this is that current obeys the rule of flowing from minus to plus, whether the voltage source is dc, such as a battery, or ac, such as a generator.

OPPOSING VOLTAGES

Cells can be connected in series (Fig. 2-13), with the total voltage output equal to the sum of the voltages of the individual cells. In a sense, this is just another way of getting

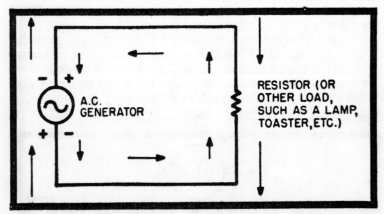

Fig. 3-13. This is essentially the same diagram as that shown in Fig. 3-12. The reversing polarity of the generator results in an alternating current.

more current control, for as voltage is increased, the amount of current flow is also increased, assuming no other circuit changes are made.

Still another cell-connecting technique is to wire them so that the voltages oppose each other. Fig. 3-14 shows a circuit arrangement of this kind. There are two batteries, wired so that the plus terminal of battery A is connected to the plus terminal of battery B. The minus terminals of each of these batteries are wired to some sort of load.

If both batteries have identical voltages, no current will flow in this circuit. The voltage of battery A opposes the voltage of battery B and since both voltages are equal, the resultant circuit voltage is zero.

Now assume battery A is 10V and battery B is 2V. The difference between these two voltages is 10 − 2 equals 8V. This circuit behaves, then, as though it had just a single battery of 8V. The direction of current flow will be decided by the battery having the larger voltage; in this case, battery A. Current will flow through this circuit in a clockwise direction.

Suppose, however, that battery A is 3V and battery B is 15V. The resultant voltage is the difference between these two voltages, or 15 − 3 equals 12V. However, since battery B has the larger voltage, it will drive current through the circuit in a counterclockwise direction.

THE CHARGING CIRCUIT

It might seem as though connecting voltage sources in such a way would be useless, but it is a technique that is

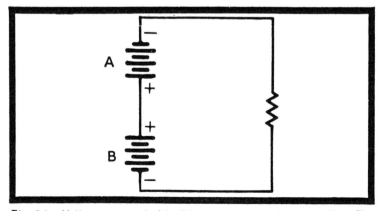

Fig. 3-14. Voltages connected in this way are in series opposition. The direction of net current flow is determined by the battery having the larger voltage.

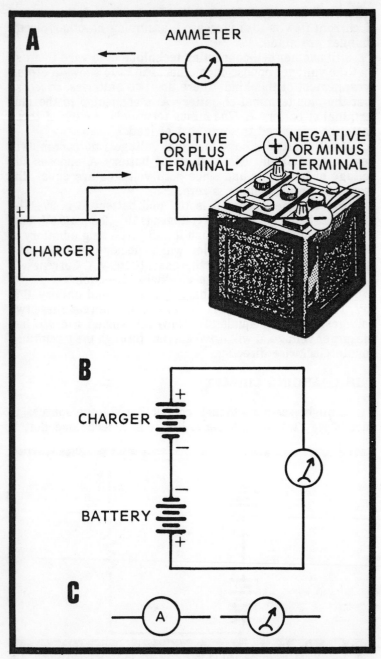

Fig. 3-15. Battery charging circuit. Pictorial (A), circuit (B). Drawing C shows two possible ammeter symbols for use in circuit diagrams.

practical and is used in certain test instruments, receiver circuits and battery chargers. As one example, consider the battery charging circuit of Fig. 3-15. In drawing A the charger has an output of about 15V while the battery being charged is 12V. The ammeter in the circuit indicates the amount of current flowing from the charger to the battery.

The circuit arrangement in A can be rearranged as in B to show that it is really the same circuit as that in Fig. 3-14. Since the charger voltage is stronger than the battery voltage, it will drive a current through the battery in a direction that is opposite to the way current normally flows from the battery. Current leaves the battery from the minus electrode when the battery delivers to a load. With the present arrangement, current is now moving into what is essentially the normal current exit. This reverse current charges the battery. In a lead-acid storage battery, for example, the lead peroxide plate and the pure sponge lead plate gradually change to lead sulphate as the battery discharges. But a requirement for a cell is that the electrodes be of different substances. During discharge, though, both plates in the lead-acid cell gradually become the same. When the battery charges, the lead sulphate changes to lead peroxide for one of the plates and to pure sponge lead for the other. Lead peroxide and sponge lead sound as though they could be identical materials, but although both contain lead, they are quite different chemically.

In Fig. 3-15, drawing A is a pictorial, drawing B is a circuit diagram arranged to correspond to Fig. 3-14. Drawing C shows two meter symbols (ammeters) used in circuit diagrams.

COMBINED AC AND DC

Sometimes a dc and an ac voltage source will be combined into a single circuuit. If you regard an ac voltage as comparable to a dc voltage with regularly changing polarity, the result will be a circuit in which the voltages of the two sources, ac and dc, either aid or oppose each other.

As an example of this sort of combined voltage operation, consider Fig. 3-16. Both voltage sources have an emf of 6V. The polarity of the battery is constant: it does not change. However, the polarity of the ac voltage does change, and so, considering only the maximum output (6V) of the generator, these two conditions exist:

$$6V + 6V = 12V$$

$$6V - 6V = 0V$$

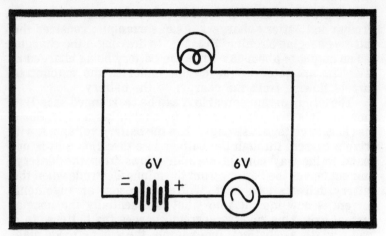

Fig. 3-16. It is possible to have ac and dc voltages in the same circuit.

When the generator lead connected to the plus terminal of the battery has a negative output, the two voltages are in series aiding, and so the total voltage output at that moment is $6 + 6 = 12V$. However, when the generator polarity reverses, the plus of the generator will be connected to the plus of the battery, and the two equal voltages will oppose. The output will then be

$$6 - 6 = 0V$$

What about the current flowing through the lamp? When the net voltage is zero there will be no current and so the lamp will not light. When the total voltage is 12V, the voltage across the lamp will be maximum and the lamp will be brilliantly lit. This circuit arrangement, then, could be used when a flashing light is needed. The number of flashes per second would depend on the ac generator; the frequency with which it would change its polarity.

The direction of this current is always in a counterclockwise direction, but in one direction only, because the generator voltage is never great enough to overcome the voltage of the battery. And so although the circuit has ac and dc voltages, the current is dc—it moves in one direction only. Note also that the current is dc even though it ranges from zero to some maximum value.

If, in the circuit of Fig. 3-16, the voltage of the ac generator is larger than the battery voltage, the current will reverse itself and an alternating current will flow. However, more current will flow in one way than in the other. Assume that the battery remains at 6V but the generator has a maximum output of 12V. When the two voltages are aiding, the total net voltage driving the current in one direction will be 12 + 6 equals 18V. When the voltage of the generator opposes the battery, the net voltage driving the current in the other direction will be 12 − 6 equals 6V. Consequently, the current will be much smaller.

By using an ac voltage in series with a dc voltage, it is possible to have a varying current flowing one way only, or to have a two-way current with unequal amounts for each direction. This is, again, a method of current control. There are circuits in radio receivers in which ac and dc voltages are used in series and in which it is important to have current control in both directions.

4 Steps Toward Current Control

A variety of different currents can flow in electrical and electronic circuits. These currents can be dc or ac, or both. Further, the alternating currents may not only have different reversal times, but there may be a number of them. And so, to prevent electrical and electronic chaos, there must be some way to direct the movement of these currents into correct paths. Further, the strength or amplitude of these currents must also be controlled. It is through current control that we are able to have such a large variety of electrical and electronic devices at our command.

There are just a few basic parts in every electrical and electronic device, although they can appear in just about every size, shape, and description. These few basic parts include tubes, semiconductors, coils, capacitors, and resistors (Fig. 4-1). Although resistors are quite simple devices, they can be made to perform a large number of important jobs.

WHAT IS A RESISTOR?

A resistor, as its name implies, opposes or resists the flow of current. It may seem strange that we generate electrical currents and then promptly produce a component for opposing them. However, we do the same thing with automobiles, providing roads which supply friction. Offhand, it would seem that no friction would be better, but then that could only be done by losing control of the autos.

BASIC CURRENT CONTROL

Fig. 4-2A shows a circuit consisting of a battery or voltage source with a wire connected directly across its terminals. Since the wire offers very little opposition to the flow of current, the battery will deliver the maximum current of which it is capable. Known as a short circuit, the effect of this connection will be to discharge the battery rapidly. The wire may become extremely hot, so much so that its insulation may

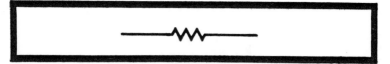

Fig. 4-1. This is the symbol for a resistor having a fixed value. The same symbol is also used to represent a load.

catch fire and set fire to adjacent parts. If a different kind of power source is used, a fuse may open. The whole point is that this massive surge of current through the connecting wire is completely out of control. The current not only does no useful work, but can cause the battery to overdischarge and become damaged. Not a good situation at all, and under certain circumstances, a dangerous one!

As a first step in controlling the current, a resistor is inserted in series, as shown in Fig. 4-2B. The amount of current that will now flow will depend on two factors: 1) the amount of voltage of the source, or battery, and 2) the amount of resistance or current opposition offered by this new part, the resistor. By controlling the amount of voltage and also by controlling the resistor, we can determine exactly how much current will flow. We can have a large current, if required, or an extremely small one. Further, with current control we can protect our voltage source against damage, we can make the current perform some work that may be required in an electrical or electronic circuit, and we have done nothing that might possibly be dangerous.

THE BASIC RESISTOR

The basic resistor (Fig. 4-3) consists of tiny granules of carbon, tightly packed together inside a tiny cylinder or tube.

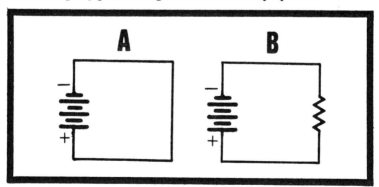

Fig. 4-2. Drawing A represents a short-circuit condition. The wire shunted across the battery represents a direct short circuit. In B, the flow of current is limited by the resistor.

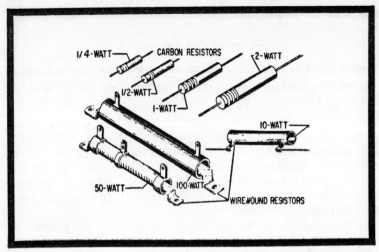

Fig. 4-3. Resistors are made in a variety of ways. Some of packed carbon granules, others are a thin layer of resistive material deposited on a nonconducting core; still others are windings of "resistance" wire.

The amount of current opposition depends on the volume of carbon granules and the way they are packed. If the volume is large, the opposition will be small. Conversely, if the volume is small, the opposition will be large. This unit, called a resistor, can be made to have any degree of current opposition, depending on the way it is manufactured.

FIXED RESISTORS

The resistor just described is known as a fixed resistor since its current opposition or electrical friction remains fixed once it is manufactured. The electrical or electronic symbol for a fixed resistor is a jagged line, as illustrated earlier in Fig. 4-1. The symbol may be drawn in any position, and while it is generally used to represent the part we call a resistor, it is also sometimes used to indicate any "load" which requires a current of electricity. It can represent a lamp, motor, radio, or television receiver. The letter R is used as an abbreviation for resistance. In some cases, in circuit diagrams, the letter R is placed directly adjacent to the symbol. But, like I (current) and E (voltage), R does not specify an actual unit of measure. The basic unit of resistance is the ohm. A resistance having a large ohmage value has a large opposition to the passage of current through it. Thus, a component having a value of 10 ohms will have ten times as much electrical friction as one having a value of 1 ohm. The Greek letter omega, capitalized, is used as an abbreviation for ohms, and so a 20-ohm resistor can be written as 20 ohms or 20 Ω. And the resistance can be expressed R equals 20 Ω.

In electricity and also in electronics, the letter k is known as a multiplier of 1,000. Hence, 1,000 ohms can be written as 1KΩ. Convention calls for the letter to be capitalized if used alone, and to be positioned right up against the value. A lowercase letter k stands for a constant, so here the k is written as a capital letter, to avoid ambiguity. A 50K resistor is one with a value of 50KΩ, or 50 kilohms, or 50,000 ohms.

CODES AND VALUES

Usually an electrical or electronic circuit may contain a number of resistors. To identify these resistors and to be able to describe their location, they are often assigned codes and values. R1 is resistor number 1; R2 is resistor number 2. This doesn't mean that R1 is better than R2, or it is doing a more important job, or is carrying a stronger current. In a diagram resistors are simply numbered from left to right and from the upper part of the diagram to the lower. Sometimes, as in Fig. 4-4, a resistor will have a value in ohms written adjacent to the symbol. Combinations of a letter, such as R, followed by a number, are called a code. The amount of resistance, in ohms, is referred to as the value. Circuit diagrams frequently indicate codes and values of resistors.

MULTIPLES OF RESISTANCE

Resistance values can range from units which are a fraction of an ohm, to millions of ohms. Common multiples of the ohm are the kilohm or thousand ohms, and the megohm, or million ohms. The abbreviation for megohm is M. A 5M

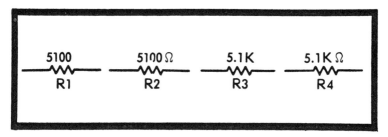

Fig. 4-4. The upper numbers show four different ways in which a resistor value can be marked adjacent to the resistance symbol (all four of the values are identical). The lower numbers show a method of coding resistors on drawings.

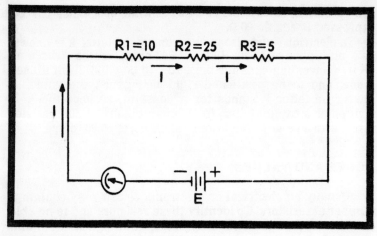

Fig. 4-5. Resistors connected in series. The same amount of current flows through each resistor.

resistor is one that has a value of 5 million ohms. Sometimes, particularly in speech, the word "megohm" is reduced to simply "meg."

CONVERSIONS

In practical work in electricity or electronics, it is essential to be able to work back and forth between ohms and kilohms. To convert from ohms to kilohms, divide ohms by 1,000. To convert from kilohms to ohms, multiply kilohms by 1,000.

THE SERIES CIRCUIT

Various combinations of resistors can be used to get very fine control of the amount of current flow. Fig. 4-5 shows three resistors connected in series, an arrangement in which the same amount of current flows through each of the resistors. The arrows indicate the direction of current flow. It isn't possible to say at just which point the current starts, any more than it is possible to indicate which part of a wheel turns first. It is convenient, though, to select a starting point, and this is usually the minus terminal of the battery, identified by a minus sign. Current flows along the conductor or wire, represented by a straight line, reaches and flows through the first resistor, R1, continues along the connecting wire to R2. The current passes through R2, through R3, and then returns to the plus terminal of the battery. From here it flows through the inside of the battery to the minus terminal. From this point, the action repeats.

OHM'S LAW

The movement of current through a resistor, and other electrical and electronic parts, isn't accidental but is controlled by natural laws. By knowing these laws and by understanding what they mean, it is possible to control and direct the movement of currents in every part of an electrical circuit.

Probably the most widely used law in electronics is Ohm's law, named after its dicoverer, George Simon Ohm. Dr. Ohm, who supplied one of the foundations for the modern sciences of electricity and electronics, lived in despair and poverty, was often out of work, and frequently ridiculed by his contemporaries. In 1817 Dr. Ohm was appointed professor of mathematics at Jesuits' College in Cologne.

Like other important laws, Ohm's law is extremely simple, but it does explain the relationship between voltage, current, and resistance. Basically, it states that the current flowing through a resistor depends on the amount of voltage. If the voltage is larger, the amount of current flow will be larger; and if smaller, the current will be smaller, assuming the amount of resistance does not change.

It is much easier to make this statement in terms of a formula:

$$\text{Voltage} = \text{current} \times \text{resistance}$$

This first form of Ohm's law states that current and resistance are directly related to voltage. The formula also indicates that the product of current and resistance, that is, current multiplied by resistance, supplies the amount of voltage.

As an example, consider a current of 2A (Fig. 4-6) flowing through a resistance having a value of 4 ohms. To determine the amount of voltage producing this current, multiply the value of current by the value of resistance, $2 \times 4 = 8$. Thus, the electrical pressure is 8V.

SIMPLIFYING OHM'S LAW

Ohm's law can be simplified by substituting letters for words. Thus, E can be used for voltage, I for current, and R for resistance. Ohm's law then looks like this:

$$E = I \times R, \text{ or } E = IR$$

83

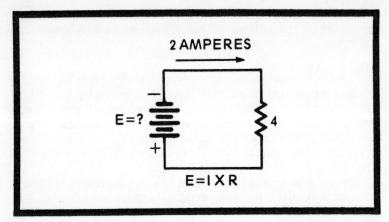

Fig. 4-6. The source voltage can be calculated by multiplying the value of the current, in amperes, by the amount of resistance, in ohms.

This formula is easy to remember because the symbol letters appear in alphabetical order from left to right.

With the help of Ohm's law, it is always possible to calculate the amount of voltage present in a circuit, provided the amount of current and the value of resistance are known. There is one requirement, however. To find the answer in volts, the current must be in amperes and the value of resistance in ohms. If the basic units of current and resistance are not supplied, then the values given must be converted into basic units.

EXAMPLES:

A current of 3A flows through a resistor of 8 ohms. The resistor is connected across a battery. What is the voltage of the battery.

To remember Ohm's law (or any other law) write it at the beginning of each problem. In this case:

$$E = I \times R$$

$$I = 3; \quad R = 8$$
$$I \times R = 3 \times 8 = 24 \quad E = 24V$$

A current of 4 mA flows through a 1.2K resistor. What is the amount of voltage supplied by a dc source delivering current to the resistor? (See Fig. 4-7.)

$$E = I \times R$$

Since the ampere is the basic unit of current and the ohm the basic unit of resistance, both values must be converted.

$$4 \text{ mA} = 0.004 \text{ ampere}$$

$$1.2\text{K} = 1,200 \text{ ohms.}$$

$$0.004 \times 1,200 = 4.8\text{V.}$$
This is the voltage of the generator.

OTHER FORMS OF OHM'S LAW

While the basic form of Ohm's law is $E = IR$, it can be modified into two other arrangements. The equal sign in the formula indicates that the quantity on the left-hand side is equal to the quantity on the right-hand side. We can manipulate this formula to our advantage provided we treat both sides of the formula equally. We can multiply or divide the left side provided we do exactly the same to the right side. As a start, write the formula:

$$E = IR$$

Now divide both sides by R and the formula becomes:

$$\frac{E}{R} = \frac{I \times R}{R}$$

Any number or letter divided by itself is equal to 1. On the right side R is divided by R and can be replaced by 1. The formula becomes:

$$\frac{E}{R} = 1 \times 1$$

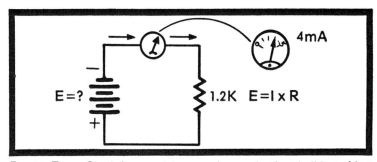

Fig. 4-7. To use Ohm's Law, resistance and current values in this problem must first be converted to basic units.

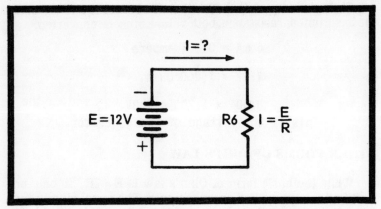

Fig. 4-8. Current can be calculated if voltage and resistance are known.

The digit 1 can be dropped and the formula can now be written as:

$$\frac{E}{R} = I, \text{ or } I = \frac{E}{R}$$

The significance of this version of Ohm's law is that it can be used to determine the amount of current flowing through a resistor if both the voltage and resistance values are known. E and R must be in basic units, and if not, must be converted.

Example:

A 6-ohm resistor is connected across a 12V battery. (See Fig. 4-8.) How much current flows through the resistor?

$$I = \frac{E}{R}, \text{ or } I = \frac{12}{6} = 2A$$

In this circuit consisting of a resistor connected across the terminals of a battery, the battery voltage is known and so is the value of the resistor. Instead of inserting an ammeter into the circuit for the measurement of the current, it is easier and faster to use Ohm's law.

Example:

A resistor having a value of 200 ohms (Fig. 4-9) is connected across a voltage source of 1.2 kV. What is the amount of current flow?

$$I = \frac{E}{R}$$

The resistance value is supplied in ohms, but the voltage must first be converted to volts. 1.2 kV = 1,200V, a figure obtained by moving the decimal point three places to the right. The problem can now be solved by "plugging" numbers into the formula.

$$I = \frac{E}{R} = \frac{1,200}{200} = 6A$$

There is still one more form of Ohm's law that can be obtained. This time both sides of the equation will be divided by I.

$$E = IR$$

Dividing both sides of the equation by I

$$\frac{E}{I} = \frac{IR}{I}$$

On the right side, I divided by I = 1 and the equation becomes:

$$\frac{E}{I} = 1R$$

The 1 can now be dropped and this final form of Ohm's law is:

$$\frac{E}{I} = R \text{ or } R = \frac{E}{I}$$

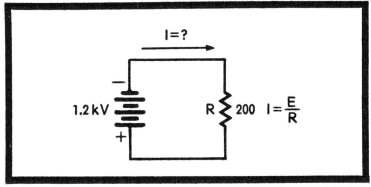

Fig. 4-9. To use Ohm's Law for solving this problem, voltage must first be converted to its basic unit.

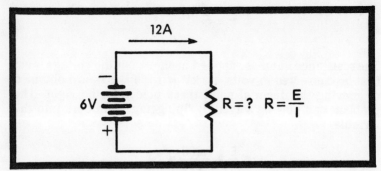

Fig. 4-10. Resistance can be calculated when current and voltage are known. While resistance can be measured with a test instrument such as an ohmmeter, it is sometimes necessary to predict the value of resistance before the circuit is constructed.

This arrangement of Ohm's law can be used to find the value of resistance in a circuit if both the voltage and the current values are known. They must, of course, both be in basic units, or converted to basic units.

Example:

What is the value of resistance (Fig. 4-10) connected across a battery when the voltage is 6V and the current has a measured value of 12A?

$$R = \frac{E}{I} = \frac{6}{12} = \frac{1}{2} \text{ ohm} = 0.5 \text{ ohm}$$

Example:

What is the value of resistance connected across a dc generator when the generator has an output emf of 120V and the measured current is 800 mA?

$$R = \frac{E}{I} \quad E = 120 \quad I = 800 \text{ mA} = 0.800A = 0.8A$$

$$R = \frac{120}{0.8} = 150 \text{ ohms}$$

The Ohm's Law Triangle

The triangle shown in Fig. 4-11 is a memory aiding device for remembering the three form's of Ohm's law. The letter E is at the top, I at the lower left and R at the lower right. To use Ohm's law for determining voltage, cover the letter E with a fingertip. The two remaining letters, adjacent to each other, are I and R. $E = I \times R$.

To learn the value of current, cover the letter I with a finger. The triangle indicates E over R, or E divided by R. Finally, to learn the amount of resistance, cover the letter R. The remaining letters are E over I, or E divided by I.

Another easy way to remember the formulas is to assign other meanings to the symbols: Think of an eagle, a river, and an Indian. The eagle (E) flies over both the Indian (I) and the river (R), so it is always placed at the top:

$$\frac{E}{R} = I \text{ and } \frac{E}{I} = R.$$

The Indian and the river are both on the ground, so

$$RI = E, \text{ or } E = IR.$$

Ohm's law applies whether a single resistor or a number of resistors in various connections are involved. Fig. 4-12 shows two series resistors, R1 and R2, connected across a 12V source. R1 has a value of 6 ohms and R2 a value of 18 ohms. The total resistance in this circuit is equal to the sum of the individual resistors. In the arrangement of a formula:

$$R = R1 + R2$$

R is the total resistance, R1 in this circuit is 6 ohms and R2 is 18 ohms.

$$R = R1 + R2 = 6 + 18 = 24 \text{ ohms.}$$

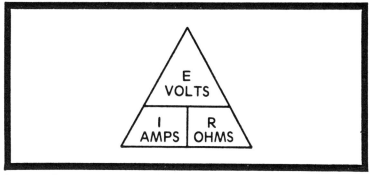

Fig. 4-11. This triangle arrangement is helpful for remembering the three forms of Ohm's law.

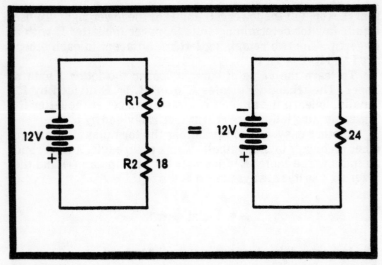

Fig. 4-12. In a series circuit the same current flows through each part. The total resistance is equal to the sum of the values of the individual resistors.

THE SERIES CIRCUIT AND OHM'S LAW

A series circuit follows Ohm's law and even helps verify it. As an example, consider the two resistors connected in series as shown in Fig. 4-14. The voltage supply is 24V; R1 is 12 ohms and R2 is 48 ohms. The total resistance R is:

$$R = 12 + 48 = 60 \text{ ohms}$$

Since this is a series circuit, the same amount of current will flow through both resistors. This current value can be calculated by using Ohm's law:

$$I = \frac{E}{R} = \frac{24}{60} = 0.4A$$

In a series circuit consisting of three resistors, the same rule still applies. To find the total amount of resistance, add the values of the individual resistors. Fig. 4-13 shows a series circuit made up of R1 = 4 ohms, R2 = 6 ohms and R3 = 10 ohms.

The same formula as that supplied earlier can be used to find the total resistance, except that now R = R1 + R2 + R3. R = 4 + 6 + 10 = 20 ohms. If there are four, or five, or more resistors in series, the total resistance is still the sum of the values of the individual resistors.

As far as Ohm's law is concerned, the two resistors in series are regarded as a single unit. However, Ohm's law can be applied to each of the resistors individually. Consider R1. Its resistance is 12 ohms and the current flowing through it, based on the problem just solved, is 0.4A. Since we have resistance

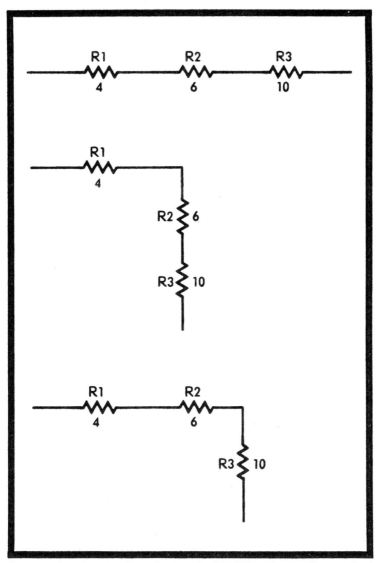

Fig. 4-13. A series circuit can be arranged in different ways. In an actual circuit, the resistors may be adjacent or can be widely separated. R1, R2, and R3 can also be interchanged without altering the total resistance.

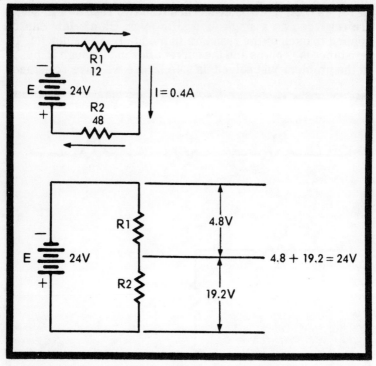

Fig. 4-14. Ohm's law applies to a circuit using resistors in series.

and current, we must have voltage, because E = IR. In the case of R1 the voltage across it is:

$$E = IR = 0.4 \times 12 = 4.8V$$

The same approach can be used for R2. Again, since we have a value of resistance and current, there must be a voltage across R2.

$$E = IR = 0.4 \times 48 = 19.2V$$

IR DROPS

The voltages across R1 and R2 are called IR drops or voltage drops. Their sum is equal to the amount of applied or source voltage. 19.2 + 4.8 = 24V.

Note that the battery voltage in this case is 24V, but that we were able to divide the voltage by means of resistors. In the example the 24V was divided into 19.2V and 4.8V. The voltage, however, could have been divided in any way we want. By using equal values of resistance for R1 and R2, the divided voltages would have been 12 and 12.

A combination of series resistors as shown in Fig. 4-14 is known as a voltage divider, but while only two resistors are shown in the illustration, a voltage divider could be made up of three, four, or even more resistors. Voltage dividers are frequently used in electronic circuits. While the diagram of series resistors shows the resistors next to each other, in an actual circuit the resistors could be widely separated.

RESISTORS IN PARALLEL

Another resistor arrangement, known as the parallel or shunt circuit, is illustrated in Fig. 4-15. To understand how different this circuit is compared to a series circuit, start at the negative terminal of the battery and follow the direction indicated by the arrows. The current follows a single path until it reaches the connection point of the resistors. Here the current divides, with some of the current flowing through R1 and the remainder flowing through R2. Unlike a series circuit, however, the current through R1 can be, and usually is, different from the current through R2. After passing through the two resistors, the two currents rejoin and flow back as a single current to the positive terminal of the battery. The current passes through the battery, emerging at the negative terminal and once again the entire process repeats.

When currents divide, as in this example, they are known as branch currents. The total current, that supplied by the

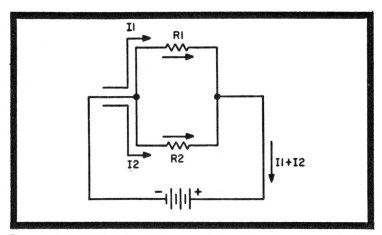

Fig. 4-15. In a parallel circuit, the current from the voltage source divides or branches. I1 and I2 are called branch currents; the current coming out of and returning to the voltage source is referred to as the line current.

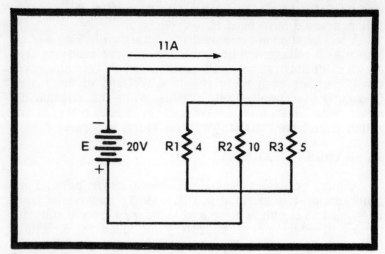

Fig. 4-16. In a parallel circuit, the voltage across each of the resistors is the same, but the current through each resistor can be different.

voltage source (a battery in this case) is sometimes called the line current. The line current divides into branch currents, with one branch current flowing through R1 and the other through R2. Although the branch currents have different values, the sum of the branch currents is equal to the line current.

The total number of branch currents depends on the number of components connected in parallel. Thus, for three parallel resistors there will be three branch currents, one for each of the parallel resistors. However, the line current will still be equal in value to the sum of the branch currents.

How to Calculate Branch Currents

Ohm's law is just as applicable to parallel circuits as it is to series circuits. As an example, Fig. 4-16 shows a parallel circuit consisting of three resistors connected to a 20-volt dc source. R1 is 4 ohms, R2 is 10 ohms and R3 is 5 ohms. To calculate the amount of current flowing through each resistor using Ohm's law:

For R1: $I = \frac{E}{R} = \frac{20}{4} = 5A$

For R2: $I = \frac{E}{R} = \frac{20}{10} = 2A$

For R3: $I = \frac{E}{R} = \frac{20}{5} = 4A$

Thus, 5A flows through R1, 2A through R2 and 4A through R3. The total current supplied by the voltage source is the sum of these branch currents, or 5 + 2 + 4 = 11A

Note that the voltage across each of the resistors is the same, a condition that is entirely different than the series circuit. In the parallel circuit, the voltage across each resistor is the same while in a series circuit each resistor can have a different amount of voltage. In a parallel circuit, the current through each resistor can be different while in a series circuit, the current through each resistor is the same.

What about the total resistance in the circuit of Fig. 4-16? That can be obtained by using Ohm's law and dividing the line current by the applied voltage. The line current is 11A and the voltage is 20V. Note that this value is less than the value of any

$$\frac{E}{I} = \frac{20}{11} = 1.8182 \text{ ohms.}$$

one of the parallel resistors.

Resistance of Resistors in Parallel

When resistors are connected in series, the total resistance is equal to the sum of the individual resistors. The arithmetic involved is easy. For resistors in parallel, the arithmetic is also simple, but takes several steps.

As an example, consider the two resistors connected in parallel as shown in Fig. 4-17. R1 is 3 ohms and R2 is 6 ohms. To find the equivalent resistance of two resistors in parallel:
1. Multiply R1 by R2.
2. Add R1 and R2.
3. Divide step 1 by step 2.

In the case of the example just supplied:
1. R1 x R2 = 3. x 6 = 18
2. R1 + R2 = 3 + 6 = 9
3. Dividing step 1 by step 2: 18 ÷ 9 = 2 ohms

Thus, a 3-ohm resistor in parallel with a 6-ohm resistor behaves as though it were a single resistor of 2 ohms. This resistance, known as the equivalent resistance, is always smaller in value than either of the parallel resistors.

Ohm's Law and the Parallel Circuit

Assume that the two resistors of Fig. 4-17 are connected to a 12V battery. R1 is 3 ohms and R2 is 6 ohms. The equivalent value has already been determined and is 2 ohms.

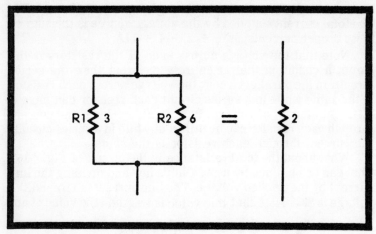

Fig. 4-17. The two resistors in parallel can be represented by a single equivalent resistor.

Now consider R1 by itself. It has a value of 3 ohms and the voltage across it is 12V. The branch current through R1 is

$$I = \frac{E}{R} = \frac{12}{9} = 4A$$

The branch current through R2 can also be calculated by Ohm's law.

$$I = \frac{E}{R} = \frac{12}{6} = 2A$$

The total current, or line current, is the sum of the branch currents. 4 + 2 = 6A. We can now find the value of R1 and R2 in parallel in two ways: we can combine R1 and R2 by the method previously described of multiplying R1 and R2 and then dividing by the sum of R1 and R2. Or, knowing the total current, we can find the equivalent resistance using Ohm's law. Here are both methods in action:

1. Combining R1 and R2, we can set up a formula:

$$R = \frac{R1 \times R2}{R1 + R2}$$

This formula is just a simplified way of describing the three steps previously described for combining parallel resistors. Substituting the values of the resistors in the formula:

$$R = \frac{3 \times 6}{3 + 6} = \frac{18}{9} = 2 \text{ ohms}$$

Since R is the equivalent resistance in the circuit, use Ohm's law to find the line current:

$$I = \frac{E}{R} = \frac{12}{2} = 6A$$

2. The other method of finding the total current is to determine the branch current through each resistor, R1 and R2, and then add these to yield the line current. The line current is 6A since one branch current is 2A and the other branch current is 4A.

Thus, in the solution of problems involving Ohm's law there is often more than one approach. Not only does this supply the technician with a choice of solutions, but gives an opportunity for checking the results. If the answers using two different solution methods are the same, then it is reasonably certain that the answers are correct.

Although the problem shown in Fig. 4-17 involves just two resistors in parallel, the same approach can be used for three or more shunt resistors.

Three Resistors in Parallel

Two resistors in parallel and three resistors in parallel aren't at all unusual in electric and electronic circuits. Sometimes even four or more resistors are shunt-connected.

There are a number of methods we can use to find the equivalent resistance of three resistors in parallel.
1. Combine two of the resistors in parallel to find the equivalent resistance. Then combine this answer with the remaining resistance to get the final answer. This requires the use of the parallel resistance formula twice.
2. Find the branch currents through each parallel resistor. Add these currents to get the total or line current and then use Ohm's law.
3. Convert each resistance into an equivalent conductance. Add the conductance and then convert the total back into resistance.
4. Use this formula:

$$R = \frac{1}{\frac{1}{R1} + \frac{1}{R2} + \frac{1}{R3}}$$

The first two of these methods have been described, but the conductance method and the formula involve some new techniques.

Conductance is the reciprocal, or inverse, of resistance, something that sounds much more complicated than it really is. The reciprocal of 10 is 1/10. The reciprocal of R is 1 divided by R. The reciprocal of any number or letter is 1 divided by that number or letter.

In the case of resistance, its reciprocal is 1/R, but this fraction is more often represented by the letter G. Thus:

$$\text{reciprocal of } R = 1/R = G$$

The conductance of a resistor also has a basic unit called the mho. This is the ohm spelled backward and since conductance is a sort of reverse resistance, this designation does make sense. If a resistor has a value of 10 ohms, its conductance, G, is 1/10 mho. To find the equivalent conductance of any resistance value, divide that value into the number 1. The result is a fractional decimal.

Example:

A resistor has a value of 2,700 ohms. What is its conductance?

$$\text{Reciprocal of } R = G, \text{ or } \frac{1}{R}. \text{ Since } R = 2,700,$$

$$\frac{1}{R} = 0.00037 \text{ mho}$$

It is sometimes convenient when three or more resistors are connected in parallel to convert the resistance values into mhos.

After making the conversion, the values of the individual conductances are added and so the three resistors, now considered as three conductors, are converted into an equivalent conductor. This equivalent conductance can then be converted to resistance again.

To simplify what may seem to be a complicated procedure:

1. Convert each individual value of resistance into conductance. Do this by dividing each resistance value into 1.
2. Add all the conductance values.
3. Divide the result of the addition into 1. The answer will be in ohms. The value will be the single equivalent value of the resistors in parallel.

Example:

What is the equivalent resistance of R1, R2, and R3 when these are wired in parallel. (See Fig. 4-18.) R1 is 100 ohms. R2 is 50 ohms and R3 is 25 ohms. Before starting the problem we know the answer must be less than 25 ohms since the equivalent resistance must always be smaller than that of the resistor having the lowest value. As a first step, convert each resistance into an equivalent conductance.

$$R1 = 100 \text{ ohms} = \frac{1}{100} \text{ mho} = 0.01 \text{ mho}$$

$$R2 = 50 \text{ ohms} = \frac{1}{50} \text{ mho} = 0.02 \text{ mho}$$

$$R3 = 25 \text{ ohms} = \frac{1}{25} \text{ mho} = 0.04 \text{ mho}$$

$$0.01 + 0.02 + 0.04 = 0.07 \text{ mho.}$$

0.07 mho represents the total conductance, G, of the three parallel resistors. To convert conductance back into resistance again, just reverse the original procedure; that is divide conductance into 1.

$$R = \frac{1}{G} \text{ and } G = \frac{1}{R}$$

$$1/0.07 = 14.2857 \text{ ohms} = 14.3\Omega \text{ approximately}$$

And so, the equivalent resistance of these three resistors in parallel is 14.2857 ohms, rounded off to 14.3 ohms.

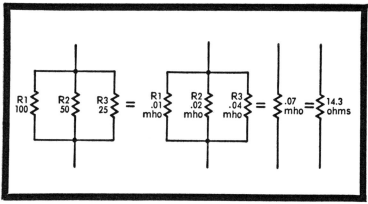

Fig. 4-18. Conductance method of finding the equivalent resistance of three resistors in parallel.

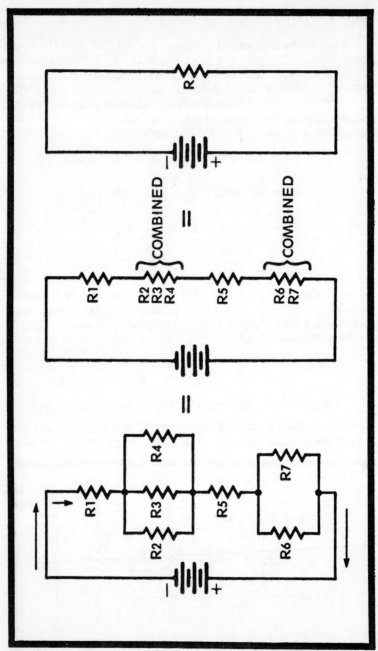

Fig. 4-19. Method for finding the total resistance, R, of a series-parallel circuit. Combine the parallel resistors into an equivalent value to convert the circuit into a series type, then add values to find the total resistance.

Formula Method. The formula method is somewhat more direct. The formula is:

$$R = \frac{1}{\frac{1}{R1} + \frac{1}{R2} + \frac{1}{R3}}$$

However, if you will examine the denominator of this somewhat large fraction you will see that each resistor is represented by its conductance, and so the formula is just a more concise way of describing the technique for converting into conductance and then back into resistance again.

$$R = \frac{1}{0.01 + 0.02 + 0.04} = \frac{1}{0.07}$$

$$= 14.2857 \text{ ohms} = 14.3 \Omega$$

THE SERIES-PARALLEL CIRCUIT

The series circuit and the parallel circuit are sometimes grouped into a combination such as the one shown in Fig. 4-19. All sorts of arrangements are possible, and that illustrated in Fig. 4-19 is just one. In this drawing, start at the minus terminal of the battery and follow the direction indicated by the arrows. The line current flows through R1. After doing so, there are three resistors in parallel, R2, R3, and R4. The current branches and so various values of current pass through these resistors. The higher the value of resistance, the smaller the amount of branch current through a particular resistor.

Following R2, R3, and R4 is resistor R5. The branch currents unite and the line current flows through this resistor. The current then branches once again as it flows through R6 and R7 in parallel. Finally, the branch currents join to form the line current which returns to the plus terminal of the battery.

In the circuit of Fig. 4-19, R2, R3, and R4 form one parallel group while R6 and R7 are another parallel group. These parallel combinations are both in series with single resistors R1 and R5.

The total equivalent resistance of this series-parallel circuit can be found on a step-by-step basis. First, combine R2, R3, and R4 into an equivalent resistance. This combined resistance will be in series with R1 and can be regarded as a single resistor. Follow the same procedure with parallel resistors R6 and R7. The result will be a series circuit. Add all the resistance values to find the single equivalent resistance, R, for the entire network.

Although the circuit of Fig. 4-19 doesn't look as simple as some of the others that have been shown, it still obeys Ohm's law. When the current flows through R1 it produces a voltage drop across it. How much? E = IR. Similarly with the following parallel network of R2, R3, and R4. Although different amounts of current flow through each of these resistors, the voltage across each of them is the same since they are in parallel.

Example:

Fig. 4-20 shows a resistive circuit consisting of R1 in series with R2 and R3 connected in parallel. This parallel network is followed by a single series resistor, R4, which in turn is wired to R5 and R6 in parallel.

R1 = 60 ohms, R2 = 40 ohms, R3 = 20 ohms; R4 = 100 ohms; R5 = 200 ohms; R6 = 50 ohms.

The voltage source is 200V.

How much current flows in this circuit? How much current flows through each resistor? What is the voltage drop across each resistor?

As a first step toward finding the total amount of current, it is necessary to find the total amount of resistance. Starting with R2 and R3.

$$R = \frac{R2 \times R3}{R2 + R3} = \frac{40 \times 20}{40 + 20} = \frac{800}{60} = 13.33 \text{ ohms}$$

Note that instead of using R1 and R2 in the formula, there is now R2 and R3. The fact that the codes were changed doesn't alter the formula.

Another technique would have been to convert each resistance in this parallel network into its equivalent conductance in mhos.

R2 = 40 ohms, so the reciprocal is 0.025 mho

R3 = 20 ohms, so the reciprocal is 0.05 mho

0.05 + 0.025 = 0.075 mho.

Converting G to R, we take the reciprocal of 0.075, or 13.333 ohms.

The equivalent resistance of R5 and R6 can be found the same way. R5 is 200 ohms and R6 is 50 ohms.

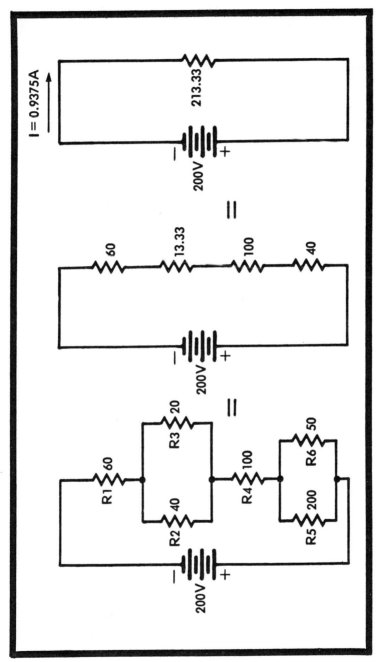

Fig. 4-20. The amount of current can be found by using Ohm's law after all the resistors have been combined into a single equivalent value.

$$R = \frac{200 \times 50}{200 + 50} = \frac{10000}{250} = 40 \text{ ohms}$$

Now that the equivalent resistance of the two parallel networks has been found, the circuit can be redrawn as shown in Fig. 4-2D. In this form it looks like an ordinary series circuit. To find the total resistance, add each resistance value. We have:

$$60 + 13.33 + 100 + 40 = 213.22 \text{ ohms.}$$

Since the source voltage is 200V, the line current can be determined by Ohm's law.

$$I = \frac{E}{R} = 200/213.33 = 0.9375 \text{ ampere}$$

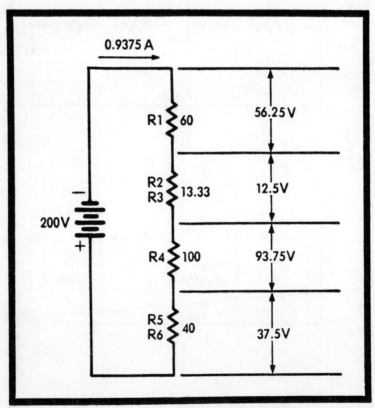

Fig. 4-21. After the parallel resistors are combined, and the total equivalent resistance is known, it is possible to calculate the IR drop across each resistor.

We are now in a position (Fig. 4-21) to calculate the voltage drop (or IR drop) across each of the resistors. In the case of R1:

$$E = I \times R1 = 0.9375 \times 60 = 56.25V$$

R2 and R3 are in parallel and so the same voltage will appear across each of these resistors. To calculate the voltage drop, though, we must use the equivalent value of R2 and R3. R2 is 40 ohms and R3 is 20 ohms. Their equivalent value, as calculated, is 13.33 ohms. The voltage across these two parallel resistors is:

$$E = I \times R = 0.9375 \times 13.33 = 12.5V$$

Since R4 is by itself, it is easy to find the voltage across it.

$$E = I \times R4 = 0.9375 \times 100 = 93.75V$$

Finally, R5 and R6 can be represented by an equivalent resistance of 40 ohms.

$$E = I \times R = 0.9375 \times 40 = 37.5V$$

To find the total voltage in this circuit, add these voltage drops.

$$\begin{array}{r} 56.25 \\ 12.5 \\ 93.75 \\ \underline{37.5} \\ 200V \end{array}$$

The above shows how the voltage is divided by the resistors.

Finding the Branch Currents

In the circuits of Figs. 4-20 and 4-21 it is easy to find the branch currents for each of the parallel circuits since the voltage drops are known. In the case of R2 and R3, use Ohm's law separately for each resistor. In the case of R2:

$$I = \frac{E}{R} = \frac{12.5}{40} = 0.3125A$$

And for R3:

$$I = \frac{E}{R} = \frac{12.5}{20} = 0.625A$$

You can add these two branch currents to find the value of line current.

```
  0.3125
  0.625
  ──────
  0.9375A
```

The same technique can be used to find the value of the branch currents for R5 and R6.

How Complicated Is It? Although combined series and parallel circuits may look complex because of the large number of components, Ohm's law is still an excellent tool for getting more information about the circuit. Briefly then, in such a circuit:

1. Combine each parallel branch into an equivalent resistance.
2. Redraw the circuit showing the parallel resistors as single equivalent resistors in series with the other resistors.
3. Add the values of all the resistors.
4. Using Ohm's law, calculate the total current.
5. Knowing the total current, determine the voltage drop across each of the resistors.
6. And finally, knowing the voltage drop across a parallel branch lets you calculate the amount of current flowing through each of the resistors in the parallel circuit.

The arithmetic involved is ordinary addition, multiplication, and division. Of course, conversions must be used if necessary to convert to basic units of resistance, current, or voltage. Electricians and electronics technicians encounter problems of this nature in their work and the ability to use Ohm's law, plus some elementary arithmetic, can supply the answers.

Effect of Changing Values

In all problems involving Ohm's law, it is assumed that the values of voltage, resistance, and current are fixed. However, in practical circuits, resistance and voltage values do change; and when this happens new arithmetic calculations, using Ohm's law, are needed.

Consider a simple circuit such as that involving a single resistor of 10 ohms placed across a 10V battery. The current flowing in this circuit will be:

$$I = \frac{E}{R}, \text{ or } 1A$$

However, if the resistance value increases, the current must decrease. Thus, if the resistance is raised to 20 ohms, that is, if its value is doubled, the current will be halved.

$$I = \frac{E}{R} = \frac{10}{20} = 0.5A$$

On the other hand, if the resistance value is reduced to half, the current will double. If the resistance is lowered to 5 ohms,

$$\frac{10}{5} = 2A$$

The same effects can be had by keeping the value of resistance fixed and changing the voltage. If the voltage is doubled, raised from 10 to 20V, the current will be doubled:

$$\frac{20}{10} = 2A$$

Similarly, if the voltage is reduced to one-half its original value, the current will be lowered by a similar amount:

$$\frac{5}{10} = 0.5A$$

We now have two factors which affect current flow: voltage and resistance. Current follows voltage directly—that is, it varies in the same way. If voltage is raised, current is increased. If voltage is lowered, current is decreased. More precisely, current varies in direct proportion to voltage.

In the case of resistance, current decreases when resistance is increased. Current also increases as resistance is decreased. Stated more exactly, current is inversely proportional to resistance.

Variable Resistors

If it is necessary to change the amount of current flowing in a circuit, doing so by connecting batteries in series or parallel can present physical problems. An easier method is to use a resistance whose value can be changed smoothly and easily. Known as variable resistors, they are commonly used

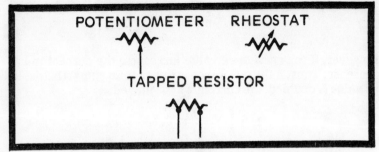

Fig. 4-22. Various types of variable resistors and their symbols.

in just about every type of circuit. The volume control on a radio receiver is a variable resistor. A dimmer for a lighting system is often a variable resistor. So is a motor speed control. The brightness and contrast controls on television sets are variable resistors. The technical name for a variable resistor is potentiometer.

Several different kinds of symbols can be used for variable resistors, as shown in Fig. 4-22. Still another type of resistor but only semivariable is the tapped unit. The resistance is changed in the sense that connections can be made to different taps. Fig. 4-23 shows the use of a variable resistor in a network consisting of series and parallel resistors. The variable resistor has a slide arm which moves across a resistive element. In Fig. 4-23 resistance is maximum when the slide arm is at the right. However, as the slide arm is moved to the left, part of the resistance element is shorted by the conductor connected to the slide arm.

THE RESISTOR COLOR CODE

If a fixed resistor is large enough, the value of resistance can be printed or stamped directly on its body. However,

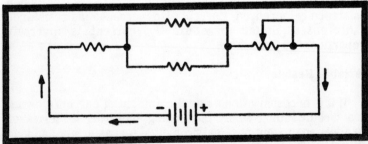

Fig. 4-23. A variable resistor can be used to change the amount of current flowing in a circuit.

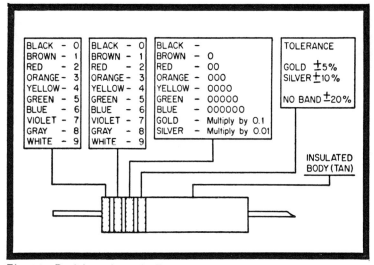

Fig. 4-24. Resistor color code. To read the resistance, hold the resistor so that the color bands are on the left.

resistors are often so small as to make this method of identification impractical. In such cases, various rings of color are used, with each color representing some number ranging from 0 to 9. This technique is known as the resistor color code and with its help you can determine the value of any resistor which is marked this way.

Fig. 4-24 shows the resistor color code, and the number corresponding to each color. To read the color code, hold the resistor so the color bands are at the left. If, as an example, the first color is red, the equivalent number is 2. This number 2 is then the first digit in the resistance value. If the next color (the second color) is brown, the corresponding digit is 1. This is the next number in the value of resistance. The third digit indicates the number of zeros following the first two numbers. If the color is orange, this means three zeros are required. And so the resistance value in this example, is 21,000 ohms, or 21K.

Example:

A fixed resistor is color-coded yellow, green, red. What is its value of resistance?

If the resistor is held so the colors are at the left, yellow is the first color, green the second color, and red the third.

yellow = 4   green = 5   red = 2 zeros = 00

The value of resistance is 4,500 ohms, or 4.5K.

The third color is sometimes called a multiplier, since adding zeros is equivalent to multiplication. Adding one zero, for example, is the same as multiplication by 10. Adding two zeros is multiplication by 100.

Tolerance

When resistors are manufactured, they may be above or below their design values. Precision resistors are those that have little or practically no deviation in resistance; these are expensive and usually found only in test instruments or in circuits where exact values are critical. For many applications, such as radio and television receivers or industrial control circuits, resistors can vary as much as 5 percent, 10 percent, or 20 percent from the amounts of resistance indicated in circuit diagrams. Resistors, other than precision types where the tolerance may be 1 percent or less, generally have tolerance-of-error values of 5 or 10 percent, although some are rated at 20 percent. (**Tolerance of error** is normally shortened to simply "tolerance.")

Percentages can be changed to decimals by moving the decimal point two places to the left. A tolerance of 20 percent is the same as 0.20; a tolerance of 10 percent the same as 0.10, and a tolerance of 5 percent equivalent to 0.05. To find the resistance deviation, multiply the value of the resistance by the decimal equivalent of the percentage tolerance. A resistor of 100 ohms, for example, having a tolerance of error of 10 percent, would be:

100 x 0.10 = 10 ohms. 100 + 10 = 110. 100 - 10 = 90.

And so a 100-ohm resistor with a 10-percent tolerance could have a true value anywhere between 90 and 110 ohms. However, this is the amount of resistance before the resistor is wired into the circuit. A resistor can change its value if heat is used in connecting it, as would be the case if a soldering iron were used, or if the resistor is mounted near heat-producing components.

Example:

What is the resistance range of a 4,700-ohm resistor having a tolerance of 5 percent.

5% = 0.05   0.05 x 4,700 = 235 ohms

While tolerance isn't ordinarily specified as a plus-minus value, this is generally understood unless otherwise indicated. In this example, then, the resistor could have any value ranging from 4,700 − 235 ohms to 4,700 + 235 ohms. 4,700 − 235 = 4,465 ohms. 4,700 + 235 = 4,935 ohms. Hence, a 4,700-ohm resistor with a 5 percent tolerance could have an actual value of any amount ranging from 4,465 ohms to as much as 4,935 ohms.

The tolerance value is the fourth color shown on the resistor, as indicated in Fig. 4-24.

INTERNAL RESISTANCE OF A CELL

The internal resistance of a cell, previously discussed in Chapter 2 and illustrated in Fig. 2-16, was not taken into consideration in the various voltage, current, and resistance problems discussed in this chapter.

Voltage and current impose different circuit requirements. Consider voltage, for example. Voltage can exist without current flow. A car parked in a garage, with all car switches turned off, still carries a 12V source; a voltmeter connected across the terminals of the battery would immediately indicate its existence. But no current flows under these circumstances, except for the very small and negligible amount of current needed to operate the voltmeter. This is an open-circuit condition.

However, when the battery is connected to a load, current must flow from the battery to and through the load. This current must also flow through the interior of the battery, from one electrode to another. The movement of current, then, requires a closed path, while voltage can be present whether the circuit is open or closed.

Inside the cell, current moves from the positive electrode to the negative electrode. The path of the current is through the electrolyte, a chemical which changes its resistance as the cell is used. Although the electrolyte doesn't resemble other conductors, such as wire or resistors, it is properly regarded as a conductor since current passes through it. And, since it is a conductor, it has a certain opposition to the passage of current, referred to as internal resistance.

Just because the current of a battery must flow through its internal resistance does not mean Ohm's law is abandoned or forgotten. The current through the interior of the battery is I, the resistance is R, and whenever a current I flows through a resistance R, there is a voltage, I x R. The voltage in this case is the internal voltage drop of the battery and must be subtracted from the terminal voltage of the battery. As an example, if a battery has a supposed terminal voltage of 12V, and an internal drop of 1V, the actual voltage available at the terminals of the battery will be 12 − 1 = 11V. Whether or not

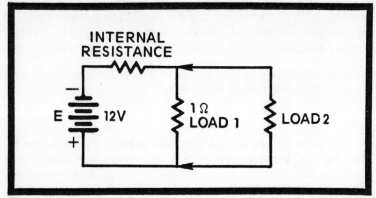

Fig. 4-25. The internal resistance of the battery is in series with the load. While the internal resistance is indicated here as a separate resistor, it is inside the battery.

this is serious depends on the battery and the load to which it is supplying current.

As the internal voltage drop of a battery becomes larger, the terminal voltage becomes smaller. But this internal voltage drop exists only because current flows through the battery. The current, however, is the current demanded by the load. Remove the load or open the circuit, and current stops flowing. But when current stops flowing outside the battery, it also does not flow through the interior of the battery. As a result there is no internal voltage drop and so the terminal voltage goes right back up. The best way, then, to measure the true terminal voltage of a battery is to do so under load, with the battery circuit connected to the device it normally operates. The two voltages of a battery, then, are its maximum or no-load voltage and its full load voltage. The full load voltage is the more meaningful, since a battery that is almost completely discharged will show full voltage across its terminals if no load is connected to it.

Internal Resistance and the Load

The internal resistance of a battery is in series with the resistance represented by the load, as shown in Fig. 4-25. Consider a 12V battery shunted by a 1-ohm resistor. This 1 ohm could represent the resistance of a motor, or a lamp, or any other device. Since the battery is fully charged and in good condition, its internal resistance is so small it can be ignored. The current flowing in this circuit, based on Ohm's law, is 12A. This is the amount of current required by the load, and under the existing circumstances the load will work properly since it is working at both rated voltage and current. If the battery has a 15A current delivering capability, there is a safety margin of 3A. Another load could be connected in parallel with the existing load, as shown in Fig. 4-25. However, this additional load should be one that will not require more than 3A. If the load does take 3A, the battery will be operating at maximum capacity.

During the time a battery delivers current its internal resistance increases from minimum at full charge to maximum when the battery is completely discharged. The internal resistance is in series with the load. If you will trace the path of current flow you will see that the same current that passes through the load also flows through the internal resistance of the battery. In the circuit of Fig. 4-25 with a single load of 1 ohm, and a 12V battery, the line current is 12A. If, however, the internal resistance of the battery rises to 1 ohm, the total circuit resistance is R1 + R2, in which R1 is the resistance of the load and R2 the internal resistance of the battery. The total circuit resistance is 1 + 1 = 2 ohms. Again, Ohm's law applies, so the current is 6A. But since the load requires 12A it will either not work at all or may work improperly. In the case of an automobile, the car may not start. In the case of a motor, the armature of the motor may not revolve. Not only has the maximum current availability been reduced by 6A, but the terminal voltage has been reduced by the amount of the internal IR drop of the battery. The battery voltage is now 12V − 6V = 6 volts.

If the battery can be recharged, the internal resistance will decrease and the terminal voltage, under load, will increase. If the battery isn't a rechargeable type, the only alternative is to replace it.

POWER

Electrical devices require both voltage and current, a combination known as power. The basic unit of power is the watt (W), and it can be calculated by using any one of a number of rules known as power laws. The simplest of the power laws is:

$$P = E \times I$$

The power used by a device or a circuit is the product of the voltage and the current. Basic units of voltage and current must be used in this formula—volts and amperes. If a problem involves multiples of these basic units, then they must be converted before using the formula. Of these four items (power, voltage, resistance, and current), only the first two are used by most electrical devices to supply information about their characteristics. An electric light bulb, for example, may be marked 75W, 120V, indicating that it is a 120-volt bulb rated at 75 watts. The resistance of the bulb and the amount of current it will take from the voltage source aren't specified, although there are some instances in which the current requirements are also listed. The current taken by a light bulb is maximum at the start and then decreases shortly thereafter since the resistance of the bulb changes. The resistance of a light bulb when cold (that is, with no current passing through it) is lower than the resistance of the same light filament after it has been working a few minutes.

Example:

An electric heater draws a current of 8A from a 115V power line. What is the amount of power being used to operate this heater?

$$P = E \times I = 115 \times 8 = 920W$$

Example:

A resistor in an electronic circuit has a current of 450 mA flowing through it. The voltage measured across the resistor is 0.165 kV. What is the amount of power being used by this resistor?

Both voltage and current in this problem must first be converted to basic units before substitution is made in the formula.

450 mA = 0.450A

0.165 kV = 165V

$$P = E \times I = 165 \times 0.450 = 74.25W$$

Other Power Laws

Just as there are three variations or different forms of Ohm's law, so too is it possible to derive other power law arrangements. The power law, as stated earlier, is:

$$P = E \times I$$

However, according to Ohm's law, E = I x R and so, since E and IR are equivalent, we can substitute IR for E in the power law.

$$P = IR \times I$$

This law can be rearranged to look like this:

$$P = I \times I \times R$$

An abbreviated way of writing I x I is I^2. The number 2 is known as an exponent and indicates that the quantity, current in this case, is to be multiplied by itself, or squared. I squared is the same as I^2 or I multiplied by I. The formula is then written as:

$$P = I^2 \times R$$

This can be further abbreviated to:

$$P = I^2R$$

Basic units of current and resistance must also be used in this formula.

Example:

A current of 3A flows through a 4-ohm resistor. How much power does the resistor use?

$$P = I^2R = 3 \times 3 \times 4 = 36W$$

Example:

A current of 250 mA flows through a resistor whose value is 1.5K. What is the power delivered to the resistor?

```
250 mA = 0.250A
1.5K = 1,500 ohms
```
$$P = I^2R = 0.25 \times 0.25 \times 1{,}500 = 93.75W$$

It is possible to develop still another power law by making use of Ohm's law once again. Starting with the basic power law:

$$P = E \times I$$

According to Ohm's law, however,

$$I = \frac{E}{R}$$

Since I and E/R are identical, one can be substituted for the other. Hence:

$$P = E \times \frac{E}{R}$$

E x E is the same as saying E squared, or E^2. The remaining power law is:

$$P = \frac{E^2}{R}$$

Note the advantage of having three power laws instead of one. If you have a problem involving E and I it is easy to calculate the power by using the formula P = E x I. However, if the problem involves current and resistance the P = E x I formula isn't much help. In that case, use P = I^2R. Similarly, if the information supplied is voltage and resistance, the other formula is the one to use.

Example:

A 25-ohm resistor in a series circuit has an emf of 50V developed across it. How much power is dissipated by this resistor? Since P is E^2 divided by R, or 50 x 50 divided by 25, the resistor uses 100 watts of power.

Example:

A 75W light bulb is connected to a 120V outlet. How much current is taken by the bulb?

$$P = E \times I$$

This formula can be transposed to read:

$$I = \frac{P}{E}$$

So with P = 75W and E = 120V, I = 0.625A, or 625 mA.

MULTIPLES OF THE WATT

There are multiples of the watt, just as there are multiples of the ampere and the ohm.

A kilowatt (kW) is equal to 1,000 watts. To convert watts to kilowatts, divide watts by 1,000 or move the decimal point to the left three places. Thus, 450W is the same as 0.450 kW. 65W can be written as 0.065 kW.

To convert from kilowatts to watts, multiply kilowatts by 1,000.

6 kW = 6W x 1000. 0.45 kW is 450W.

A megawatt (MW) is equal to a million watts. To convert from watts to megawatts, divide watts by 1,000,000 or move the decimal point to the left six places. 045,375W is equal to 0.045375 MW. 6W is the same as 0.000006 MW.

For the use of formulas involving watts, it is usually necessary to convert from kilowatts or megawatts to watts. Sometimes, though, depending on the problem, it may be necessary to move back and forth between kilowatts and megawatts, particularly in electrical problems.

To convert from kilowatts to megawatts, divide kilowatts by 1,000. To convert megawatts to kilowatts, multiply megawatts by 1,000. Thus, 645 kW = 0.645 MW. Similarly, 35 MW = 35,000 kW. Although division and multiplication are indicated in each of these examples, actually the decimal point was moved back and forth as required.

While the kilowatt and megawatt are multiples of the watt, this unit also has two submultiples, that is, values smaller than a watt. There are a number of submultiple units, but the most common (and useful) of these are the milliwatt and the microwatt. A milliwatt is a thousandth of a watt and is abbreviated mW; a microwatt is a millionth of a watt, abbreviated uW. The microwatt may seem as though it is too small a unit to be practical, yet radio signals across receiving antennas are usually measured in terms of microwatts. And components in electronic circuits are frequently specified in terms of milliwatts.

To convert from watts to milliwatts, multiply watts by a thousand. To convert from milliwatts to watts, divide milliwatts by one thousand. To convert from watts to microwatts, multiply watts by one million. To convert microwatts to watts, divide microwatts by 1,000,000. A milliwatt is a thousand times as large as a microwatt. To convert milliwatts to microwatts, multiply milliwatts by 1,000. And to convert microwatts to milliwatts, divide microwatts by 1,000. (See Fig. 4-26.)

1 watt	1,000 milliwatts
1 watt	1,000,000 microwatts
1 watt	0.001 kilowatt
1 watt	0.000001 megawatt
1 kilowatt	1,000 watts
1 kilowatt	0.001 megawatt
1 megawatt	1,000,000 watts
1 megawatt	1,000 kilowatts
1 milliwatt	0.001 watt
1 milliwatt	1,000 microwatts
1 microwatt	0.000001 watt
1 microwatt	0.001 milliwatt

Fig. 4-26. Power multiple and submultiple conversion.

POWER RATINGS OF APPLIANCES

Applicances are frequently rated in terms of the amount of power they use. Appliances that convert electric energy to heat energy usually have the highest power ratings and, of course, are the units that require substantial currents. The list of Fig. 4-27 isn't absolute—you will find many variations. Thus, the first appliance listed is shown as a lamp, rated at 60W. This simply means that this lamp uses a 60W electric bulb. It is also possible to have lamps that use 100 or 200W, or just 40W; however, these power ratings will supply some indication of how appliances compare with each other in terms of power requirements.

WIRE & CABLE

Although wires are among the most common types of conductors, most any metal, liquid or gas that permits the

APPLIANCE	POWER RATING IN WATTS
Lamp	60
Shaver	10
Electric Heater-Single Room	1000
Sunlamp	275
Electric Range	12,500
Hot water heater	3,000
Refrigerator	250
Portable mixer	100
Coffee maker	660
Table fan	75
Waffle iron	660
Toaster	1,200
Vacuum cleaner	400
Washing machine	350
Hand iron	1,000
Radio	50
Television receiver	300

Fig. 4-27. This table shows how different electrical appliances consume different power values. Note that when energy is converted to heat, the power requirements start climbing.

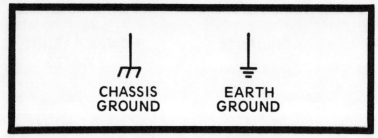

Fig. 4-28. The ground symbol is used in electrical and electronic diagrams to indicate a common connection. The rake indicates chassis ground; diminishing horizontal lines indicate earth ground.

movement of electrons through it is a conductor. Salt water is an electrical conductor; and so is a water solution of an acid or a base. Mercury vapor, a gas, is a conductor. A metal chassis supporting electronic parts is a conductor and is generally used as such. An antenna and its connecting lead-in is a conductor. So is the earth.

Like any other substance, the earth has a certain amount of conductivity, but this is quite variable, depending on particular geographic conditions. A salt, marshy area is a better conductor than a desert region. In the early days of radio, a connection to earth or ground was important because the earth formed part of the receiving system of the radio. Today, ground means a common connection. In electronic equipment, ground is sometimes a metal chassis on which the various parts are mounted. The chassis is used as a common connecting point. Sometimes a length of bare wire is used instead of a metal chassis for a ground or common linking point. The ground symbol is conveniently used in circuit diagrams since it helps simplify them, minimizing the number of conductors that must be represented in the diagram.

Sometimes, though, the word "ground" is used to mean earth ground. A ground wire on an electrical appliance, for example, is intended to be connected to the steel box of an electrical outlet. The box is grounded to earth by a waterpipe or other conductor actually buried in the earth.

In schematics, the two types of ground are distinguished by different symbols, as shown in Fig. 4-28.

WIRE TABLE

Just as there are cells capable of delivering larger or smaller amounts of current, so too are wires designated by the currents they can safely carry. A wire can be forced to carry more current than the amount for which it has been designed, but in this process electrical energy is converted to heat energy. The heat increases the resistance of the wire and this effect reduces the current flow. If, however, the electrical pressure is increased further, the larger current flow will cause the wire to glow with heat and the wire may ultimately burn out. This is the theory behind the use of fuses. Fuses are designed to carry a designated amount of current, but are made of a metal with a low melting point. An increase in current beyond the rated amount will cause the temperature of the metal to increase, melting it and opening the circuit.

The current carrying capacity of a wire is determined by its cross-sectional area. Of course the longer a wire is, the greater its resistance and so for long wires the length is a factor to consider. The cross-sectional area of wire is specified by its gage number, ranging from 0000, the thickest wire, to 46, the thinnest. Fig. 4-29 is a table showing the relationships of bare copper wire. Known as the American Wire Gage (AWG), the table usually supplies information about the resistance of the wire per unit length and its resistance at various temperatures. Fig. 4-30 supplies an indication of the actual cross-sectional area of some selected wires.

Cross-Sectional Area of Wire

A wire can be round or it can be square. In the case of a square, the cross-sectional area can be found by multiplying any two sides. If a square is 2 in., this means every side is 2 in. The area of such a square would be 2 x 2 = 4 sq in. Area dimensions of wires, however, aren't specified in inches, but in thousandths of an inch. A mil is a thousandth of an inch and so a mil is equal to 0.001 in. It is much more convenient to work with mils than with inches, since wire sizes in mils are in whole numbers. They could also be in inches, but then the inches would be in the form of fractional decimals. A square wire that is 2 mils would have 2 x 2 = 4 sq mils area. In terms of square inches this would be 0.002 x 0.002 = 0.000004 sq in.

For round wires the cross-sectional area can be easily calculated, for it is equal to the diameter squared. The diameter, as shown in Fig. 4-31, is the longest unbroken single straight line that can be drawn inside the circle, touching the perimeter of the circle. Hence, the cross-sectional area of a circular wire is d^2 where d is the diameter. A circular wire having a diameter of 5 mils will have a cross-sectional area of 5 x 5 = 25 circular mils.

The wire table shown in Fig. 4-29 supplies the diameter and the cross-sectional area of copper wire in circular mils. To

AWG	Diameter in mils	Area in circular mils	Ohms per 1000 feet
0000	460	211,600	0.0490
000	409.6	167,772	0.0618
00	364.8	133,079	0.0779
0	324.9	105.560	0.0983
1	289.3	83,694	0.1239
2	257.6	66,358	0.1563
3	229.4	52,624	0.1970
4	204.3	41,738	0.2485
5	181.9	33,088	0.3133
6	162.0	26,244	0.3951
7	144.3	20,822	0.4982
8	128.5	16,512	0.6282
9	114.4	13,087	0.7921
10	101.9	10,383	0.9989
11	90.74	8283	1.260
12	80.81	6530	1.588
13	71.96	5178	2.003
14	64.08	4107	2.525
15	57.07	3257	3.184
16	50.82	2583	4.016
17	45.26	2048	5.064
18	40.30	1624	6.385
19	35.89	1288	8.051
20	31.96	1021	10.15
21	28.46	810.0	12.80
22	25.35	642.6	16.14
23	22.57	509.4	20.36
24	20.10	404.0	25.67

find, for example, the diameter of No. 18 wire (18 AWG), locate 18 in the AWG number column at the left. Move to the right and in the first adjacent column you will find the diameter of the wire in mils. Still moving to the right, the next column supplies the cross-sectional area in circular mils.

Resistance and Temperature

The resistance of a wire varies with temperature. As the temperature increases, the resistance of the wire also increases. However, if the resistance increases, the effect is to decrease the amount of current flowing through the wire. This

AWG	Diameter in mils	Area in circular mils	Ohms per 1000 feet
25	17.90	320.4	32.37
26	15.94	254.1	40.81
27	14.20	201.6	51.47
28	12.64	159.8	64.90
29	11.26	126.3	81.83
30	10.03	100.6	103.2
31	8.928	79.71	130.1
32	7.950	63.20	164.1
33	7.080	50.13	206.9
34	6.305	39.75	260.9
35	5.615	31.53	329.0
36	5.000	25.00	414.8
37	4.453	19.83	523.1
38	3.965	15.72	659.8
39	3.531	12.47	831.8
40	3.145	9.89	1049
41	2.80	7.84	1323
42	2.49	6.20	1673
43	2.22	4.93	2104
44	1.97	3.88	2672
45	1.76	3.10	3348
46	1.57	2.46	4207

Fig. 4-29. Bare copper wire table. American Wire Gage (AWG). Resistance values are measured at 68° Fahrenheit (20° Celsius).

may or may not be serious, depending on the length of the wire, the amount of resistance change, and the work the wire is doing. For small amounts of current, involving less than an ampere, resistance changes in wire due to temperature are of little consequence. However, the larger the amount of current, the more serious a resistance change will be. Examine 18-gage wire. This is a wire commonly used for connecting electrical circuits. Locate 18 in the AWG column and then move to the right. The resistance of a thousand feet of this wire at 68° Fahrenheit is 6.385 ohms. This resistance increases to 7.51 ohms, little more than a 1-ohm difference, when the temperature rises to 149°F. However, this is for 1,000 ft of wire. If

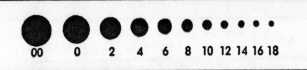

Fig. 4-30. Relative cross-sectional areas of wires of gage numbers 00 to 18.

you are connecting two electrical parts and are using about 1 ft of wire, the resistance increase would be one thousandth of one ohm or 0.001 ohm for this particular wire. Obviously, this small amount of resistance increase will hardly affect the amount of current flow.

Square Mils vs Circular Mils

Fig. 4-31 shows the cross-sectional areas of two wires, one circular, the other square. The circular wire has a diameter of 1 mil and the square wire has four sides, each 1 mil. If the circular mil area is placed on top of the square mil area it is easy to see which is greater. The circular mil area is only 0.7854 as large. Or, the square mil is 1.2732 larger than a comparable circular mil.

To convert circular mils to equivalent square mils, multiply circular mils by 0.7854. If a wire has a cross-sectional area of 250 circular mils, then its equivalent square mil area is 250 x 0.7854 = 196.35 sq mils. In effect, then, a copper wire having a cross-sectional area of 196.35 sq mils will carry as much current as a wire having a cross-sectional area of 250

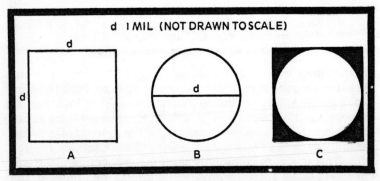

Fig. 4-31. The cross-sectional area of a square wire (A) is equal to any two sides multiplied (d^2). The cross-sectional area of a round wire is equal to the diameter (d) multiplied by itself (d^2). The diameter in B is equal to the length of a side of the square in A. Drawing C shows that under these conditions, a square mil has more area than a circular mil.

circular mils. Conversely, a wire with a cross-sectional area of 125 square mils will carry as much current as a wire with a cross-sectional area of 125 x 1.2732 = 159.15 circular mils. To check the arithmetic, multiply 159.15 by 0.7854. 159.15 x 0.7854 = 125 mils.

Stranded Cable

Wire conductors are sometimes paired or twisted together to form a cable. This has two advantages. It results in a stronger wire and it increases the current carrying capacity. The wires of the cable are usually of the same diameter and so, to find the total cross-sectional area of a cable, add the individual areas. A cable consisting of 7 strands of 18-gage wire would have a total diameter of 7 x 40.3 mils = 282.1 mils.

Wire Types

Just as there are many different types of batteries and numerous kinds of resistors, so too will you find wire in just about every conceivable style. Fig. 4-32 shows just a few of the many that are available. Drawing A of Fig. 4-32 illustrates a pair of stranded-conductor wires. The conductors making up the strands may be bare copper covered with a first layer of cotton thread insulation. Surrounding the cotton thread is another insulator such as rubber, or possibly a synthetic rubber such as Neoprene, or some plastic material. Further insulation is supplied by more cotton braid followed by an outer cotton-braid sheath. Although drawing A shows stranded conductors being used, you will also find solid wire conductors.

Drawing B shows a group of conductors having a certain number of twists per linear foot. This not only strengthens the cable but puts each individual wire in closer pressure contact with its neighboring wire. The wires are covered with felted asbestos, a material which has better fire-resistant qualities than cotton. The outer sheath is made of braided asbestos. The asbestos reduces possibility of fire in the event the wire is short-circuited, and prevents heat transfer for such high-current loads as heaters, electric irons, and the like.

Wires and cables are sometimes exposed so they are subjected to a variety of weather conditions. Where cables are situated so they can get wet, they are protected by watertight jackets of lead or rubber. Drawing C in Fig. 4-32 shows a triple conductor, stranded cable, possibly insulated with synthetic rubber. The filler is used to help supply a rounded shape and to

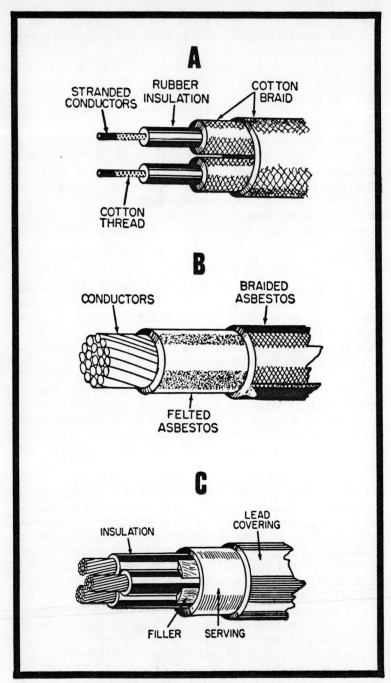

Fig. 4-32. Stranded conductor arrangements.

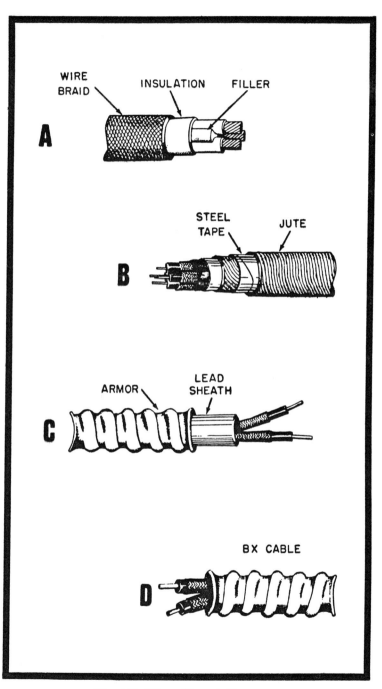

Fig. 4-33. Wire with metallic armor.

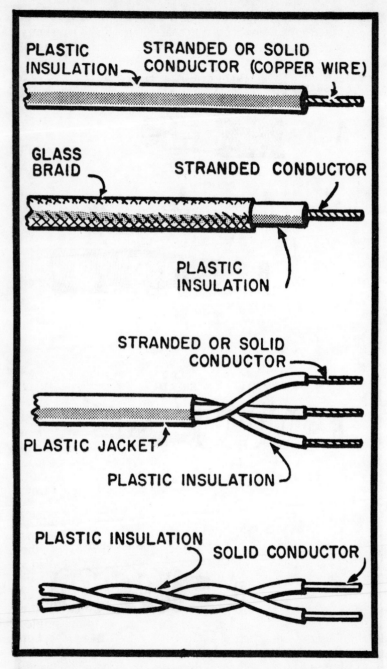

Fig. 4-34. This type of wire is designed to carry lighter current loads than the wires shown in Fig. 4-30 and Fig. 4-31.

support the serving. The serving is an insulating inner sheath which is then covered with a lead outersheath. When wire may become subject to oil splashing, synthetic rubber is often used, since it is much more impervious to oil than natural rubber.

Metallic armor is often used to cover wires for home and factory installations, as shown in Fig. 4-33. The wire braid type, A, is used where the wire must be bent around corners. The wire braid forming the outer sheath is made of steel, copper, or aluminum. If the wire carries alternating currents, the sheath may be grounded to act as a shield against unwanted radiation from the cables.

Another type of metallic armor is steel tape, in drawing B. The tape is covered with an insulating layer. Drawings C and D show two types of BX cable, one with a lead sheath (C). BX cable is often used for house wiring, although in some installations pipe or conduit is now being used instead.

Fig. 4-34 shows wires that do not carry such heavy currents as the metallic shielded types. Such wires are more simply insulated, with the insulation some type of plastic or glass braid. This is the kind of wire you will find in radio and television receivers or in appliances which have limited current loads.

SWITCHES

Switches are used to close or open circuits, permitting current to flow or not to flow. Like most electrical and electronic parts, they are available in a tremendous number of sizes, shapes, and styles (Fig. 4-35). The basic switch shown in Fig. 4-36 consists of a blade or pole with some kind of handle to permit the pole to be moved. The switch shown in Fig. 4-36 is used to open (break) or close (make) a single circuit. Technically, the switch is known as a single-pole, single-throw type, abbreviated as spst.

Sometimes it is necessary to make or break two circuits at the same time. This can be achieved by combining two spst switches with some sort of nonconducting link which forces the two switches to work in unison. A switch of this kind is called a double-pole, single-throw, abbreviated dpst (Fig. 4-37).

Still another switch is one that can be used to break or make either one circuit or another. This variation of the single-pole, single-throw switch (Fig. 4-38) is called the single-pole, double-throw (spdt). Just one pole is used, as before, but now the pole can make or break two circuits, but not at the same time. With this switch, either one circuit can be closed, or the other, or both may be opened.

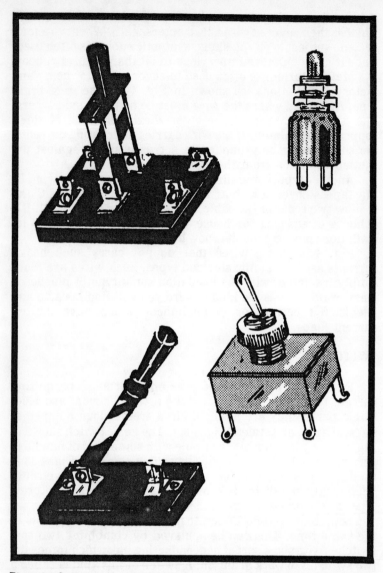

Fig. 4-35. Switches are available in a large number of sizes, shapes and styles.

Another switch variation is the double-pole, double-throw (dpdt), which is (Fig. 4-39) similar to a pair of spdt switched in tandem. This switch can be used to break or make two pairs of circuits. The poles are mechanically connected by an insulated arm and so each switch is electrically independent of the other.

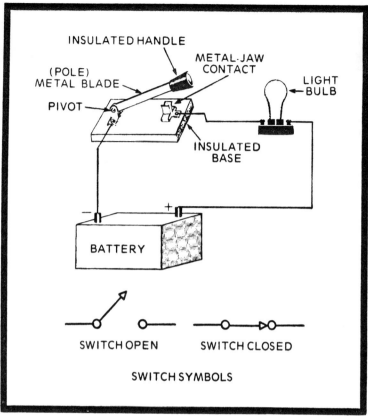

Fig. 4-36. The switch shown here is a single-pole, single-throw type.

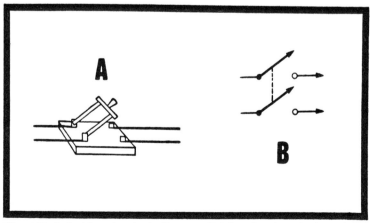

Fig. 4-37. A double-pole, single-throw switch has four terminals and can open or close a pair of conductors. Pictorial (A); symbol (B).

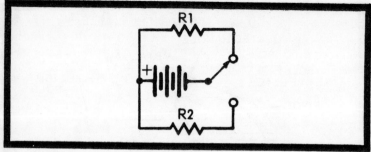

Fig. 4-38. Circuit showing symbol of single-pole, double-throw (spdt) switch. When the pole of the switch is moved to the left, R1 is in the circuit. When the switch blade is moved to the right, R2 is in the circuit.

FUSES

The flow of current through a conductor such as a wire or a resistor always involves the transformation of electrical energy into heat energy. In a properly designed circuit the amount of heat developed may be so small as not to be noticeable. In other instances, as in the case of electric light bulbs, the heat may be considerable. The material through which the current flows must be capable of dissipating the heat, throwing it off or getting rid of it in some way. If not, the

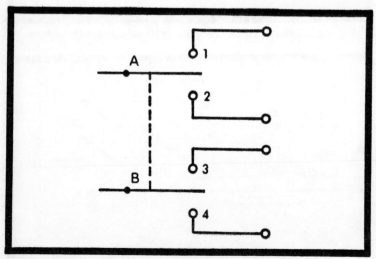

Fig. 4-39. Double-pole, double-throw relay contacts. The two poles, A and B, are connected mechanically and move together. When the first pole touches contact one, the second pole touches contact 3.

heat accumulates, and eventually the component may start to burn—nature's own way of dissipating heat rather rapidly.

The development of heat is accelerated when a higher-than-required current flows in a circuit. This can be caused by poor design or through the failure of the component in some respect. Whatever the reason may be, circuits are often protected against excessive currents by fuses. A fuse is a metal resistor with two characteristics: low resistance and a low melting point. Fuses are made so that a relatively small amount of heat will make them change from a metal to a liquid to a gas—rather rapidly.

Fuses are wired in series with the circuit they are designed to protect. Fig. 4-40 shows a pictorial of four 1.5V cells in series, supplying a total of 6V. The resistor in the circuit is 29 ohms while the fuse has a resistance of 1 ohm. The total circuit resistance is 30 ohms, assuming negligible resistance in the connecting wires and the blade of the switch. Based on Ohm's law, the current in this circuit will be 200 mA. The fuse in Fig. 4-40 is rated at 0.5A (500 mA) meaning that it is designed to safely carry this amount of current.

If the resistor should become shorted by a wire, the circuit resistance would decrease from 30 ohms to 1 ohm. Using Ohm's law again, the current would increase substantially. $I = E$ divided by R, or 6A. Since this amount of current greatly exceeds the capacity of the fuse, it would promptly open, thus protecting the battery and also the connecting wires.

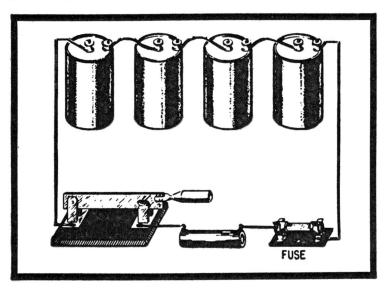

Fig. 4-40. Fuse is wired in series with current-carrying line.

FROM DC TO AC

Up to now the emphasis on the circuits that have been presented has been on a dc voltage source. While batteries are an extremely important part of our technology, they represent only one voltage type. And so the next chapter will introduce a new kind of voltage concept—the generation or production of currents for immediate use, rather than for storage and use, as in the case of batteries.

How Voltages Are Generated

There are a number of ways of generating a voltage, and a battery is just one of them. The battery, a dc source, is a chemical factory for it converts chemical energy into electrical energy. However, the first method ever developed for producing a voltage consisted of rubbing some material, a nonconductor, with fur or cloth. This technique could be called the friction method; and while some friction machines produced spectacular sparks of electricity they couldn't be put to practical use. The development of the battery finally put the science of electricity on the right path, for the battery supplied a steady, dependable source of electricity that could be put to work.

The battery, though, wasn't, and isn't, enough. It has to be recharged and, in its early form, was bulky and supplied only a very low voltage. There was no practical way, in the early days of battery development, of taking the battery voltage and increasing it. Fortunately, another method was at hand.

FROM A STRAIGHT WIRE INTO A COIL

There is nothing too remarkable about an ordinary straight wire, but when the wire is wound into the form of a coil, some of its properties, weak in the form of a straight wire, become highly emphasized. An electric current consists of a flow of electrons and these moving electrons are each surrounded by a tiny magnetic field. A wire carrying a current, then, acts like a magnet because of the movement of the electrons through it. And, just like a permanent magnet, a wire carrying a current has a north pole and a south pole. It isn't the copper wire that creates the magnetic field but the electrons flowing through it. If the wire is made of aluminum or zinc, the same magnetic field exists and the wire, whether aluminum or zinc, also has north and south poles.

If the current through the wire is increased, the strength of the wire magnet also increases. Conversely, if the current is reduced, the strength of the wire magnet is also reduced. And,

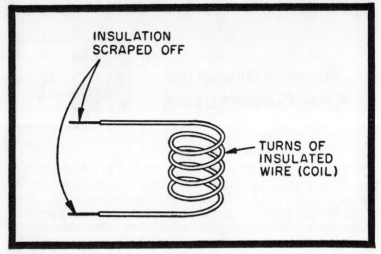

Fig. 5-1. A stronger magnetic field can be obtained by winding wire into the form of a coil.

if the current is stopped, the wire no longer shows any evidence of being a magnet. A wire carrying a current is a temporary magnet since its behavior as a magnet depends entirely on the current flowing through it.

Sending a stronger current through a wire to increase the magnetic field around the wire has its practical limitations. As the current increases, the heat produced by the moving current may cause the wire to glow and burn out. However, the magnetic strength can be increased by winding the wire into the form of a coil (Fig. 5-1). Since there are now turns of wire adjacent to each other, the magnetism around adjacent turns supplies a reinforcing action, so the result is a more concentrated or stronger total magnetic field.

Types of Coils

Since winding a wire into the form of a coil increases its magnetic strength, obviously the best way to produce a very strong magnet is to wind as many turns as possible. Unfortunately, the coil then becomes very long and cumbersome. A reasonable solution is to wind another layer of wire right on top of the first layer (Fig. 5-2). With this technique, it is possible to produce coils having several hundred turns, but still occupying comparatively little space.

To keep the turns of wire from shorting against each other, the wire is covered with an insulating material such as silk, cotton, nylon, or enamel.

Fig. 5-2. To conserve space, coils are often wound in layers.

Coil Names

There are a variety of coils designed for particular use in electrical or electronic circuits. Generally, coils are named according to the way they are constructed, or for some physical characteristic, or for the way in which they are used. Thus a solenoid is a tubular coil for the production of a magnetic field. A solenoid is also known as a helix. When a current passes through a solenoid, the south pole is at the end at which the current flows clockwise to an observer facing it.

Coils are also known as single-layer, double-layer, or triple layer, depending on the number of layers of turns. They

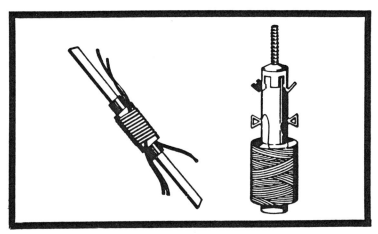

Fig. 5-3. Iron can be used as a core material for coils.

137

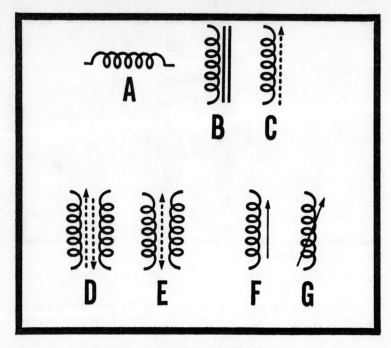

Fig. 5-4. Various coil symbols.

may also be identified by the type of core around which they are wound. If the core is completely open, the coil may be referred to as an air-core coil. Sometimes iron in the form of thin sheets or powdered iron is used inside the coil form (Fig. 5-3). In this case the coil is known as a ferrite-core type. In some coils, the core is movable; in others it is fixed.

Coil Symbols

There are various kinds of coil symbols, as shown in Fig. 5-4. The basic symbol appears in drawing A. The letter L is used to identify coils when they are used in a circuit and, as in the case of resistors, they can be numerically coded as L1, L2, etc.

A pair of solid straight lines adjacent to the coil (B) indicates the coil uses a solid core—solid only in the sense that the core consists of thin sheets or laminations.

In some cases, the core consists of powdered iron, compressed with a binder material, and taking the form of a "slug." The slug may be slotted to permit adjustment by an insulating tool somewhat resembling a screwdriver, but made of a nonconducting material. A slotted screw is sometimes inserted into the core, called a polyiron slug, permitting the screw end to protrude from the coil form, while the slug remains inside. The dashed line in C indicates a polyiron slug.

A pair of coils may be placed adjacent to each other, wound on the same form. Each coil will have its own polyiron slug, individually adjustable. The arrowheads at top and bottom indicate that one coil is tuned from the top of the coil form, the other from the bottom. An alternative method of drawing this symbol is shown in E. Drawing F represents any coil with a tunable slug.

All of the coils in A to F have a fixed number of turns. While it isn't easy to change the number of turns of a coil, the coil can be tapped so various numbers of turns can be used, the coil is variable in this sense. Drawing G shows the symbol for a variable coil. A coil is also referred to as an inductor or sometimes as an inductance.

Magnetic Strength

The strength of the magnetic field around a coil (Fig. 5-5) can be increased by sending more current through the coil, by increasing the number of turns of wire, or both. Each of these techniques has its limitations. Sending too much current through the coil can damage it and, in any event, the current must be supplied by some voltage source, such as a battery. The greater the amount of current required by the coil, the greater the current drain on the battery.

Increasing the number of turns has several disadvantages. As the number of turns is increased, the coil

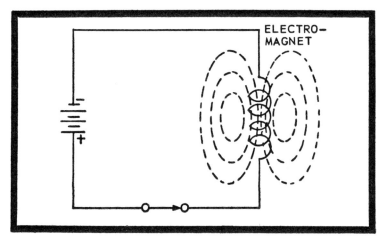

Fig. 5-5. A current flowing through a coil surrounds the coil with its electromagnetic field. The field extends into the space around the coil.

becomes bulkier and so may not fit in the space reserved for it in a circuit. As the number of turns is increased, the overall resistance of the coil also increases. But more resistance means smaller current flow and so having more turns to get a stronger magnet can mean a smaller current, which would result in a weaker magnet. Consequently, these two methods of increasing the magnetic strength of a coil often represent a compromise.

The Ampere-Turn. Since the strength of the magnetic force around a coil is determined by current and number of turns, a unit, the ampere-turn, is often used for designating magnetic strength. The ampere-turns of a coil is equal to the number of turns multiplied by the amount of current, in amperes, flowing through it. A coil having 100 turns and 2 amperes of current will have a magnetic strength of 100 x 2A = 200 ampere-turns. However, a coil having 400 turns will have the same magnetic strength if a current of 0.5A flows through it, since 400 x 0.5 = 200 ampere-turns.

There is still another unit, the gilbert, or G, for indicating magnetic strength. One ampere, turn is the same as 1.257G. To convert ampere-turns to gilberts, multiply ampere-turns by 1.257. A coil having a magnetic force of 200 ampere-turns would be identical with a coil having a magnetic force of 200 x 1.257 = 251.4G.

Reluctance

When a current flows through a conductor the opposition to the movement of current is called resistance. The lines of flux which exist between the north and south poles of a magnet also encounter opposition. Thus, it is much easier for lines of flux to pass through a substance such as soft iron rather than air. The opposition to magnetic lines of force is called reluctance. Because iron has so much less reluctance than air it is sometimes used as a core material and may even be used to surround the coil completely. And, because the reluctance or opposition to magnetic lines is so much less when iron is used, the magnet is stronger. Thus, there are three ways of making a strong magnet: 1) use a stronger current; 2) use more turns of wire for the coil; and 3) use an iron core.

Permeability

There are two ways of thinking of the effect of a wire on a current flowing through it. One is to consider the wire as supplying a path having a certain conductivity. The other is to regard the wire as having a specific quantity of resistance. These two terms are both used to describe the wire, but they are reciprocals: the greater the resistance, the less the conductivity. It's like thinking of a bottle that is half full. It can also be described as being half empty.

Similarly, we can regard magnetic lines as encountering opposition between the north and south poles of the magnet, an effect we call reluctance. However, instead of describing the magnetic path in terms of opposition, it can also be pictured in terms of ease or permeability. Permeability is the ratio of the magnetic conducting ability of iron compared to that of air, and is the reciprocal of reluctance. If one is increased, the other is decreased, and vice versa. Reluctance is comparable to the resistance of a wire. Permeability is similar to its conductivity.

Polarity of a Coil's Magnetic Field

When current flows through a coil, the magnetic field, or lines of magnetic flux which surround the coil, extend outward from the coil at right angles to the turns of wire. One end of the coil is the north pole; the other end, south. Fig. 5-6 shows there is a relationship between the direction of flow of current and the poles. The arrows on the coil indicate the way in which the current moves. Should the current reverse, the magnetic poles will also reverse.

As in the case of permanent magnets, each line of flux is continuous, unbroken, and extends from the axis of the coil out into space. Since the inside of the coil has a smaller cubic volume than the space around it, and since all magnetic lines

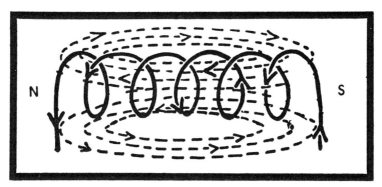

Fig. 5-6. The polarity of the poles of an electromagnet depends on the direction of current flow.

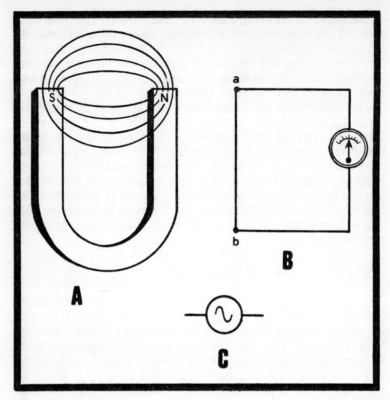

Fig. 5-7. Elementary equipment for generating a voltage. Drawing C is the symbol for a generator.

pass through it, the magnetic flux density is greater inside the coil than out. When an iron core is placed inside the coil, the reluctance is considerably decreased, and because the magnetic lines find it so much easier to go through the iron, there are a larger number of lines. However, these same lines must also exist in the open space around the coil where the reluctance is so much higher. This is a limiting factor and tends to restrict the number of lines.

THE INDUCED VOLTAGE

A commonly used method of having a current flow through a conductor, such as a wire, is to connect the wire to a voltage source. The voltage source could be a battery and, as long as the battery is charged, a current will continue to flow. In other words, since the battery can supply an electromotive force, or emf, there will be a current through the wire.

There is still another technique for developing or generating a voltage instead of using a battery. Fig. 5-7 shows the method. The drawing illustrates a horseshoe magnet (A) with lines of magnetic flux existing in the open space between the two adjacent poles of the magnet. The other piece of equipment (B) consists of a straight length of copper wire (a–b) to which is connected to a very sensitive current-reading meter. When the wire is moved between the poles of the magnet at right angles to the lines of flux, the pointer of the meter will move slightly, indicating a flow of current through the wire.

When a conductor, such as a length of wire, moves across or "cuts" the magnetic lines of force, electrons in the wire are crowded toward one end of the wire. Since the electrons are no longer evenly distributed along the length of the wire, there is a voltage across the wire. A voltage, as described earlier, always exists when there is a difference in electron quantity between two reference areas (Fig. 5-8). In the case of the wire, electrons move toward one end of the wire, but in so doing, the

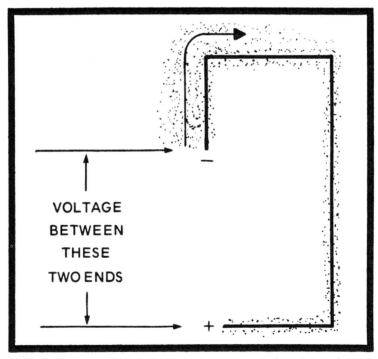

Fig. 5-8. A concentration of electrons along one end of a wire produces a voltage across that wire (with the polarity shown). Electrons will migrate from the crowded toward the less-crowded end.

143

other end of the wire is deprived of them. The voltage that is produced across the ends of the wire is called an induced voltage and the migration of electrons toward one end of the wire is referred to as an induced current.

Unlike a battery, however, the induced voltage and induced current are temporary. As soon as the wire stops moving, or as soon as it stops cutting the magnetic lines of force, the electrons redistribute themselves uniformly along the length of the wire. There is no longer an induced voltage and no longer an induced current.

While the induced voltage and current cease when the wire stops moving, the voltage and current can be obtained again by moving the wire once more, provided the wire remains in the flux field of the magnet and provided it moves at right angles and not parallel to the lines.

It isn't necessary to move the wire. Instead, the wire can remain fixed in position and the magnet moved. If the magnet is moved, the lines of magnetic flux are cut by the wire, and a voltage is induced. If the wire is moved, the wire cuts the lines of magnetic flux, and once again a voltage is induced. But if both wire and magnet are moved at the same speed and in the same direction, there will be no induced voltage. The requirement for an induced voltage is that the conductor must move across lines of magnetic flux.

THE BASIC GENERATOR

The voltage produced by the experiment described in Fig. 5-7 is extremely small, so small that it would need to be measured by a highly sensitive instrument. A more practical arrangement would be to take a long length of wire and wind it into the form of a coil, as shown in Fig. 5-9. The circuit arrangement is basically the same as that shown in Fig. 5-7 except that the straight wire is now wound on a coil form and the horseshoe magnet has been discarded in favor of a bar magnet. The galvanometer (Fig. 5-10) is a zero-center-reading sensitive current meter. When no current flows through the meter, the pointer of the meter rests at zero, but zero is at the center of the scale. When the pointer moves to the right it indicates current is flowing in a particular direction. However, when the meter pointer moves to the left of zero, it also indicates current is flowing, but in the opposite direction.

When the bar magnet is moved into the coil, the meter pointer moves to the right of the zero mark on the galvanometer. If the bottom end of the coil form is closed the bar magnet will only be able to travel a certain distance and

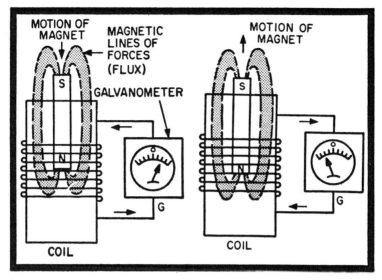

Fig. 5-9. A voltage is induced across the coil as long as the magnet is moved in and out of the coil. The magnetic lines of flux, not shown, extend across coil and surround it.

will stop. When the magnet stops moving, the pointer of the galvanometer will return to center zero. This indicates that all current flow has stopped.

If, at this time, the magnet is pulled out of the coil, the galvanometer pointer will move once again, but this time toward the left of center zero, as shown in the drawing.

The way the galvanometer pointer moves is significant. When the pointer is at the right of zero, current flows in one direction through the coil. But when the pointer is at the left of

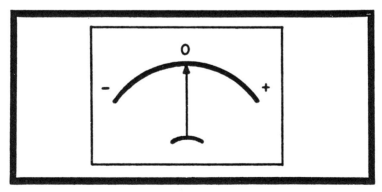

Fig. 5-10. The galvanometer is a sensitive current-reading meter. Zero current flow is indicated when the meter pointer rests at center.

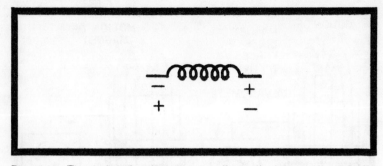

Fig. 5-11. The induced voltage across the coil changes its polarity regularly. The number of times the polarity changes depends on how fast the magnet moves in and out of the coil.

zero, we have evidence that current is now flowing in the opposite direction.

As in the case of the straight wire, the voltage across the coil is an induced voltage. When the magnet is first thrust into the open core space of the coil, the lines of flux of the magnet are pushed across the turns of wire of the coil. Electrons in the coil move to one end, making that end negative with respect to the opposite part of the coil. The meter, however, and its connecting wires supply a path for these crowded electrons and so they flow through this connecting link in an effort to reach the uncrowded end of the coil. It is this movement of electrons that produces the motion toward the right of the galvanometer pointer. However, when the magnet is pulled out, electrons in the coil rush to the opposite end. Once again this electron difference between the ends of the coil is equivalent to a voltage or electrical pressure. And once again the electrons flow through the connecting wires and the meter, but this time in a direction completely opposite to the way the current flowed originally.

The effect here is similar to that given in an earlier illustration showing wires being transposed to a battery. The difference is that in the case of the moving magnet, the change of polarity is accomplished automatically. The fact that the polarity of an induced voltage changes regularly is shown in Fig. 5-11. The polarity of the voltage across the coil is indicated by plus and minus signs, just as they are on a battery.

Alternating Voltage and Current

The current flowing through the coil of Fig. 5-9 changes or alternates its direction, depending on the direction of movement of the magnet into the coil. For this reason the current is called alternating, and is referred to as ac. The polarity of the induced coil voltage also changes every time the moving magnet changes its direction, and so is properly called alternating, and is also referred to as ac.

The AC Generator

The moving magnet, coil, and galvanometer of Fig. 5-9 is a useful laboratory experiment to demonstrate the generation of an alternating voltage and current, but a more practical model is shown in Fig. 5-12. In this instance the magnet is fixed in position while the coil is made to move. The moving coil, known as an armature, is made to revolve through the magnetic lines of flux between the north and south poles of the magnet by a crank handle. In a commercial generator instead of this laboratory model the armature would be mounted on a shaft which would be turned by some device such as a motor, steam engine, or turbine.

Since the armature rotates, there must be some way of permitting the current flowing in the armature coil to move to an external circuit, represented here by the galvanometer and its connecting wires. Mounted on the same shaft with the armature are a pair of collector rings, also known as slip rings. One end of the armature is connected to one of these rings and the other end to the second collector ring. Resting against the slip rings are a pair of brushes. These brushes make good contact with the rotating slip rings.

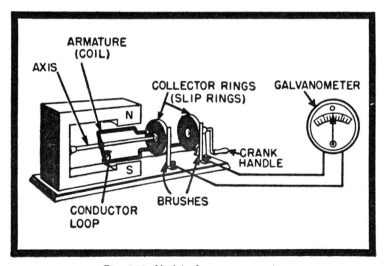

Fig. 5-12. Model of an ac generator.

And now, when the armature coil is rotated, a voltage is induced across it and the current flows from the armature coil, through the slip rings, through the brushes to the galvanometer. In a commercial type of generator, the brushes would be made of carbon or possibly copper, both of which are conductors. Instead of a galvanometer, the two leads connected to the brushes might be connected to a bank of lamps which would light due to the generator current passing through their filaments.

The faster the armature is rotated, the more often does the polarity of the voltage induced across the armature coil change and the more often does the direction of current flow change. The number of times in a second that voltage changes polarity (and current reverses direction) is referred to as frequency. A pair of reversals—that is, current flowing first one way then the other—is called a cycle. The current delivered to your home by your local power station has 60 such cycles per second. The unit of measure that means cycles per second is "hertz," abbreviated "Hz"; and, like other electrical units, it has its multiples (mega and kilo).

The DC Generator

The current from the ac generator described in Fig. 5-12 can be changed to dc by a simple modification—the substitution of a commutator, which consists of a slip ring split into two equal halves separated by some insulating material. The commutator allows the current to flow in one direction only out of the generator. In the dc generator, as in the ac version, the current flowing in the armature coil still reverses itself. However, in the dc generator the two split halves of the commutator are so arranged and positioned that current can flow out of the armature coil in one direction only.

MOTORS vs GENERATORS

A generator is a device for converting mechanical energy to electrical energy. Some type of mechanical device is necessary for rotating the armature so that electrical energy can be produced. This is true whether the generator is an ac or dc output type.

A motor is the opposite of a generator. Electrical energy is put into a motor, but the output is a rotating shaft which can perform useful mechanical work. Both motors and generators are referred to as dynamos. A dynamo, then, is an electromechanical device which can convert mechanical energy to electrical energy, or which can take electrical energy and convert it to mechanical energy. While the word dynamo is sometimes used to describe a particular type of machine, such as a dc generator, it isn't restricted to this one application.

All generators can be classified under two main headings, depending on the kind of output they have: that is, whether they supply ac or dc. Ac generators are sometimes also known as alternators. As a general rule, in dc generators the magnets are fixed in position while the armature coil revolves. In ac generators, however, the armature is kept stationary while the magnet revolves. Actually, other than design factors, it makes no difference, since the whole purpose is to get relative motion between a conductor, such as a coil of wire, and a magnetic field.

THE CONCEPT OF INDUCTANCE

While it may be strange to think of it as such, a moving electric current is a magnet. Visualize a current flowing through a wire and also imagine that the current keeps changing its strength. The magnetic field surrounding the moving current will also change and will be strong when the current is strong and weak when the current is weak.

If such a varying current is sent through a wire we have all the conditions previously explained for producing an induced voltage. The varying current is accompanied by a varying magnetic field. This varying magnetic field cuts across the wire carrying the current and so we have the strange situation in which a current of varying strength induces a voltage across the wire through which it flows. The voltage produced by moving a magnet in and out of a coil is called an induced voltage; that produced by a current flowing through the coil is referred to as a self-induced voltage.

Fig. 5-13 shows a coil connected to a generator. Because the generator periodically changes its polarity, the current will sweep back and forth through the coil. But as the current flows through the coil, it will induce a voltage across it. We now have two voltages in the circuit of Fig. 5-13. One of these is the generator voltage while the other is the self-induced voltage appearing across the coil. Both of these voltages are ac, and their polarities are such that they constantly oppose each other. Thus, when the top end of the generator in Fig. 5-13 is plus, so is the top end of the coil connected to it. And when the polarity of the generator changes and the top end becomes minus, so too does the top end of the coil.

Opposing Voltages

Fig. 5-14 (drawing A) shows a 12V battery connected across a 2-ohm resistor. With this circuit arrangement, the

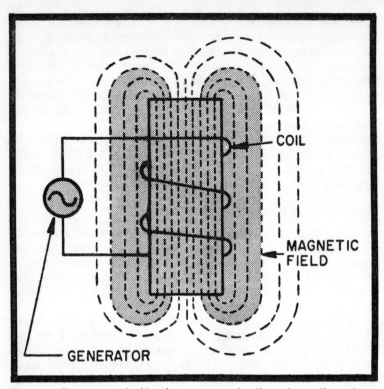

Fig. 5-13. The current flowing from a generator through a coil carries a magnetic field. This magnetic field induces a voltage across the turns of the coil.

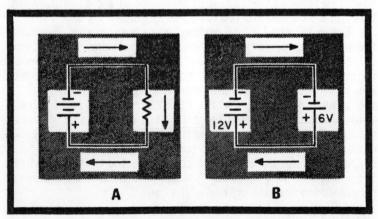

Fig. 5-14. In drawing A, the only opposition to the flow of current is the resistor. In B, the 6V battery opposes the flow of current from the 12V battery.

current (calculated from Ohm's law) is 6A. If we now change this circuit by adding another battery, as shown in drawing B, the current will be reduced even though we now have two batteries instead of one. To see why this should be so, consider the way the second battery has been connected. Its plus terminal is wired to the plus terminal of the original battery. And so, while one battery tries to drive the current in one direction, the second battery will try to force the current the other way. Voltages connected in this way are in series opposition; the voltages oppose each other. Since the original battery is 12V and the new battery is 6V, the total effective voltage in the circuit is the difference between the two: that is, 12V − 6V = 6V.

Effect of Self-Induced EMF

In Fig. 5-15 an ac generator has an output of 10V. The current flowing through the coil connected to the alternator has a self-induced emf of 8V. But the self-induced emf opposes the voltage of the alternator, and so the total effective voltage

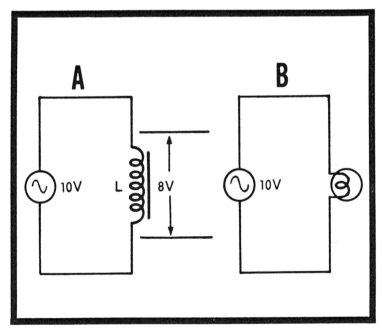

Fig. 5-15. In drawing A the self-induced emf of the coil opposes the generator voltage. The net voltage is the generator voltage minus the voltage induced across the coil. In drawing B the circuit voltage is the full 10V of the generator.

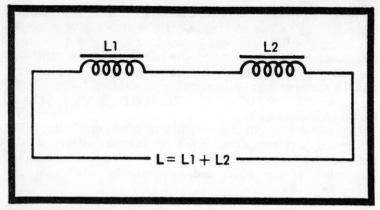

Fig. 5-16. The total inductance of two coils in series when their magnetic fields do not interact is the sum of the inductance values of the individual coils.

in this circuit is just $10 - 8 = 2V$. This is the voltage that must be used in calculations for determining the amount of current flowing through the coil.

In drawing B of Fig. 5-15, the coil has been replaced by a lamp. The lamp, however, does not contain a coil but a single wire known as a filament. As a result the magnetic field of the current has no effect on it and there is no self-induced emf. Hence the voltage in the circuit is the full voltage of the generator and in this case is 10V.

The voltage produced across a coil by a current flowing through it is sometimes called a counter emf because it counters or opposes the applied or generator voltage. The property of a coil which permits it to produce a counter emf is called inductance. The basic unit of inductance is the henry (H). In electronics, the submultiples of the unit (micro and milli) are quite common.

Inductance of a Coil

A straight wire has very little inductance, just another way of saying that a straight wire is incapable of having more than a small induced counter emf. When the wire is wound into the form of a coil, the inductance becomes greater; more turns means larger inductance. The inductance is also increased when an iron core is substituted for the air core of the coil. If a varying current flows through a coil, a certain amount of counter emf will be induced in the coil by the current flowing through it. This counter emf can be increased just by putting an iron core into the coil form.

COILS IN SERIES

Two coils can be connected in series, as shown in Fig. 5-16. In this circuit the coils are widely separated and so there is no interaction between the magnetic fields surrounding the two coils. The total inductance of the two coils is equal to the sum of the individual inductances. In the formula, **L** means total inductance, while L1 represents coil number 1 and L2 is coil number 2.

$$L = L1 + L2$$

Sometimes the coils are sufficiently close so that the magnetic fields (when current flows through the coils) interact and aid each other. The effect of the magnetic field of one coil (carrying current) on another is called mutual inductance, abbreviated with a capital M (which is unfortunate, since M also denotes "millions"). In terms of a formula, the total inductance:

$$L = L1 + L2 + 2M$$

As in the case of inductance, the mutual inductance is in henrys (H), millihenrys (mH) or microhenrys (uH), depending on its value. (Note that henrys is not spelled with an "ie" ending when written as a plural.)

The current path through a pair of coils can be arranged as in Fig. 5-17A or as in Fig. 5-17B. The path of current in drawing B is such that the magnetic fields oppose each other, while in drawing A the magnetic fields aid. When the magnetic fields oppose, the formula becomes:

$$L = L1 + L2 - 2M$$

Example:

A coil having an inductance of 1.2H is connected in series aiding with a coil having an inductance of 340 mH. The mutual inductance is 4 mH. What is the total value of inductance?

In this problem, two of the given values are in millihenrys so the easiest step to take is to convert 1.2 henrys to millihenrys.

$$1.2 \text{ henrys} = 1.2 \times 1000 = 1200 \text{ millihenrys}$$

The mutual inductance is 4 mH. L1 = 1200 mH and L2 = 340 mH.

$$L = L1 + L2 + 2M = 1200 + 340 + 2 \times 4 = 1200 + 340 + 8 = 1548 \text{ mH}.$$

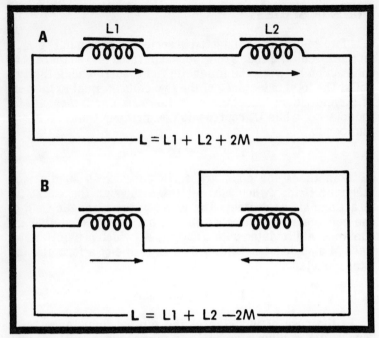

Fig. 5-17. The magnetic fields of two coils aid each other when the coils are sufficiently close and current flows through them in the same direction (A). The magnetic fields oppose when current directions are not the same (B).

In this same problem, turning one of the coils around so that the magnetic fields would oppose, or "buck" each other, would result in a total inductance of:

L = L1 + L2 - 2M = 1200 + 340 - 8 = 1532 mH.

Coils in Parallel and Series-Parallel

Coils can be wired in parallel as well as in series, as shown in Fig. 5-18A. Sometimes series-parallel arrangements are used as illustrated in Fig. 5-18B.

The Coil and Direct Current

To get a self-induced emf across a coil, the current flowing through the coil must vary, such as the current produced by an ac generator. Consider the case, however, of a coil connected across a battery, as shown in Fig. 5-19. The battery voltage does not change its polarity and almost im-

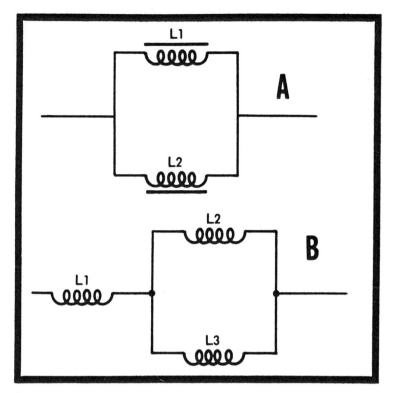

Fig. 5-18. Coils can be wired in series, parallel, or series-parallel. Drawing A shows a pair of iron-core coils in parallel. Drawing B illustrates three air-core coils in series parallel.

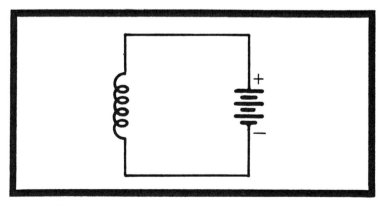

Fig. 5-19. No counter emf is produced across the coil since the current is accompanied by an unchanging magnetic field. The only limitation to the amount of current flowing in this circuit is the resistance of the coil and the connecting leads.

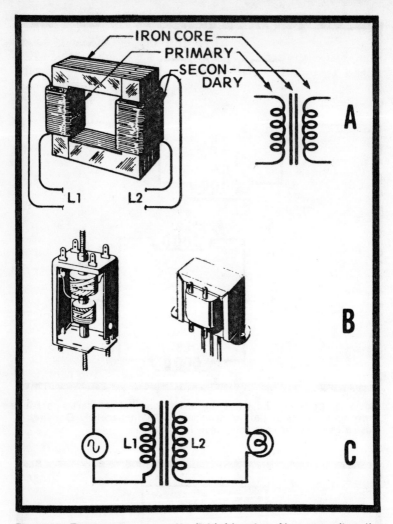

Fig. 5-20. The varying magnetic field (drawing A) surrounding the primary also covers the secondary and induces a voltage across it. The iron core, indicated by two parallel lines, helps in transferring the lines of magnetic force. Drawing B shows two of the many types of transformers. Drawing C shows the transformer connections.

mediately after the switch is closed, the current through the coil is steady. There is a magnetic field surrounding the coil produced by the current flowing through it. But this current is unchanging, and as a result its magnetic field is also unchanging. Since the magnetic field does not cut across or through the wires of the coil, there is no self-induced emf. There is no counter voltage produced by the direct current.

Consequently, the only opposition to the flow of current is the resistance of the coil. However, if a 10V alternator is substituted for the 10V battery, the amount of current flowing will be less, depending on the number of turns of wire of the coil and the frequency with which the current changes its direction.

THE TRANSFORMER

Fig. 5-20 shows a pair of coils placed fairly close to each other. There is no physical connection between the coils. The first coil, or primary, can be connected to an ac generator or any ac source. The second coil could be wired to a lamp. A combination of coils arranged in this way is called a transformer, with L1 referred to as the primary winding, or more simply the primary, and L2 the secondary. L1 plus its connecting wires and the ac source is referred to as the primary circuit while L2, its connecting wires and lamp is the secondary circuit.

There is now a varying current flowing in the primary circuit. The amount of current will depend on the source voltage, the resistance of the connecting wires, the resistance of primary coil L1, and the amount of counter emf of L1. There will, however, be a varying magnetic field around L1 because the current flowing through it is a changing one. This magnetic field, though, is large enough to cover all of L2 and because the magnetic field is a changing one, will move back and forth across the turns of wire of the secondary coil. As a result a voltage will be induced across the secondary. Since the secondary is a closed circuit, a current will flow in this circuit, and if the amount of secondary current and voltage are correct, the lamp (known as the load) will light.

The One-to-One Transformer

The primary and the secondary of the transformer shown in Fig. 5-21 can have the same number of turns, in which case the transformer is known as a one-to-one, or a 1:1, type. The two straight lines between the primary and secondary indicate that both primary and secondary coils are wound on a core. The core, in this case, consists of thin sheets of laminated metal whose effect is to increase the inductance of both coils. The laminations also supply a low-reluctance path for the magnetic field of the primary so it may more easily reach over and surround the secondary.

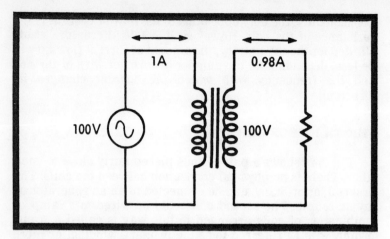

Fig. 5-21. In a 1:1 transformer, the voltage across the secondary is the same as the primary voltage, since both windings have the same number of turns.

With a 1:1 transformer the amount of voltage induced across the secondary winding is equal to the voltage of the primary winding. If the primary voltage is 100V, the voltage appearing across the secondary is also 100V. Offhand, it might seem as though we have a 100-percent-efficient device here, for the voltage is the same on both sides of the transformer. However, we must consider current as well as voltage.

In the circuit of Fig. 5-21 a current of 1A flows in the primary winding. Since the voltage across the primary is 100V, the power developed in the primary circuit is $P = E \times I = 100 \times 1 = 100W$. If the transformer is 98 percent efficient, then the electric power delivered to the secondary will be 0.98 times 100, or 98W. The remaining 2W will be lost in the transformer itself, mostly in the form of heat.

Since the voltage across the secondary is the same as the primary voltage, but the output power is 2 percent less, it follows that the current delivered by the secondary must be smaller than the current in the primary circuit. The current can be calculated from the formula used earlier:

$$P = E \times I$$

by transposing, the formula can be changed to:

$$I = \frac{P}{E}$$

The power is 98W and the emf is 100V. Hence, the current drain is 0.98A.

Stepdown and Stepup Transformers

The stepdown transformer is one in which the secondary winding has fewer turns than the primary. And, as you might expect, a stepup transformer (Fig. 5-22) is one whose secondary has more turns than the primary.

The ratio of the number of secondary turns to the number of primary turns is called the turns ratio. We can use the letters N_s to represent secondary turns and N_p to indicate primary turns. The turns ratio, then is:

$$N_s : N_p$$

The turns ratio of a 1 to 1 transformer is 1 since the number of secondary turns and the number of primary turns are identical. For a stepup transformer, the turns ratio is a number greater than 1. A stepup transformer having 100 primary turns and 500 secondary turns would have a turns ratio of 5:1, or 5. A stepdown transformer, however, has a turns ratio that is a fraction. If a stepdown transformer has 1000 primary turns and 500 secondary turns, the turns ratio, $N_S : N_p = 1:2$, or one-half. The turns ratio for a stepdown transformer, then, is

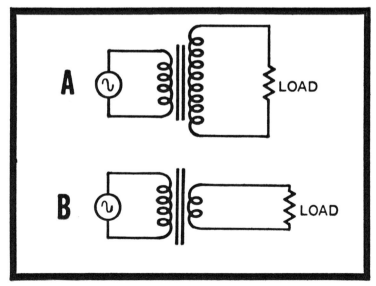

Fig. 5-22. Stepup transformer (A) and stepdown type (B).

always a fraction; for a stepup transformer it is always a whole number, while it is unity or 1 for a 1:1 transformer.

The turns ratio determines the amount of voltage appearing across the secondary of the transformer and the amount of current flowing through it. Secondary voltage and current can be calculated by using some electronic formulas:

Secondary voltage = turns ratio x primary voltage

To find the turns ratio, divide the number of secondary turns by the number of primary turns.

Example:

A transformer has 500 primary turns and 100 secondary turns. The primary is connected to a source which delivers 50V ac. How much voltage appears across the secondary? The turns ratio $N_s:N_p = 1:5$, or one-fifth, which is 10V.

The words stepup and stepdown, then, refer to voltage. A stepup transformer is one that steps up or increases voltage. A stepdown transformer does just the opposite: it steps down the voltage.

While a stepup transformer increases the voltage so there is more voltage across the secondary than across the primary, it reduces the current by the same factor. If a stepup transformer increases the voltage across the secondary by a factor of 5, for example, it will reduce the current by the same amount. If a stepup transformer has a primary of 100V and a 5:1 turns ratio, then the secondary will be 100 x 5 = 500V. If the primary current is 1A, the secondary current will be 1 divided by 5, or 0.2A.

For a stepdown transformer, however, the results are exactly the opposite. The voltage is stepped down, but the current is stepped up by the same factor. If a transformer has a primary of 200V and 4A and a turns ratio of 1:2 (0.5), the secondary voltage will be 200 x 0.5 = 100V, but the secondary current will be 4 x 2:1, or 8A, again assuming an efficiency of 100 percent.

Power In, Power Out

A transformer is a very efficient device and it isn't at all uncommon for some of them to have efficiencies very close to 100 percent. A transformer is purely electronic and has no moving parts. As a result, it is often convenient to ignore transformer losses and to assume that the power delivered to the primary winding by an ac source is the same as that appearing in the secondary circuit of the transformer (Fig. 5-23). Thus, if the primary power of a transformer is 100W, we can reasonably assume the secondary power of the same transformer is also 100W.

In the previous example, the primary voltage was described as 200V and the primary current as 4A. Since power is the product of voltage and current, then the primary power is:

$$P = E \times I$$

$$P = 200 \times 4 = 800W$$

In the example given, the turns ratio was specified as 1:2 for this stepdown transformer. The secondary voltage was calculated to be 100V and the secondary current, according to the arithmetic used, was 8A.

$$P = 100 \times 8 = 800W$$

Transformer Efficiency

It is possible to have a transformer whose efficiency must be taken into consideration. With any transformer, however, the voltage is not affected by the efficiency, no matter what the efficiency may be. A stepup transformer having a 5:1 turns ratio, and a primary voltage of 100V will have 500V across its secondary winding, regardless of how efficient the transformer is. The amount of voltage appearing across the secondary winding of a transformer is fixed by the turns ratio, whether the transformer is a stepup or a stepdown type. What is affected is the secondary current. The less efficient the transformer, the smaller the amount of current flowing in the secondary winding, whether the transformer is a stepup or a stepdown type.

Example:

A transformer has a primary of 50V and 2A. The transformer is a 2:1 stepup type and has twice as many turns on its secondary winding as on its primary winding. Assuming 100 percent efficiency, what is the amount of secondary voltage? What is the amount of secondary current? What happens to the secondary voltage and secondary current if the efficiency of the transformer is 80 percent?

Under conditions of 100 percent efficiency:
Secondary voltage = primary voltage times turns ratio = 50 x 2:1 = 100V
Secondary current = primary current times inverse of turns ratio (turns ratio upside down) = 2 x 1:2 = 1A

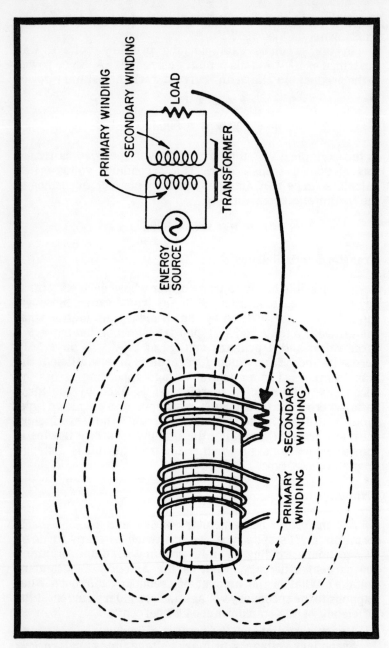

Fig. 5-23. While a generator is referred to as a voltage source, it delivers voltage and current, hence is a power or electrical energy source. The energy delivered to the primary by the generator is transferred via the magnetic field to the secondary winding.

Proof:

Primary power = secondary power for 100 percent efficiency
Primary power = 50 x 2 = 100W
Secondary power = 100 x 1 = 100W

If the efficiency of the transformer is only 80 percent, then the primary power and secondary power are no longer the same. The secondary power = 80 percent = 0.80 x primary power = 0.80 x 100 = 80 watts. The secondary voltage, however, is 100V. The secondary current can be calculated from:

$$I = P/E = 80/100 = 0.8A$$

Proof: 0.8 x 100 = 80W

A simpler way of doing the same problem would be to calculate the secondary current and then to multiply the answer by the efficiency.
Secondary current = 1A for 100 percent efficiency.
Secondary current = 1 x 0.80 = 0.80A for 80 percent efficiency.

Note that the secondary voltage is not affected by the efficiency factor in any way.

Loaded vs Unloaded Transformers

A transformer is an excellent device with no moving parts for transferring electrical energy from a source, such as a generator, to a load, such as a lamp, radio, toaster, etc. However, it is entirely possible for the load to become disconnected from the secondary. The primary may still be wired to the electrical energy source, but since it no longer is required to deliver energy to the secondary, its own energy demands decrease sharply. Since the primary is still connected to the voltage source, there is still a voltage across the primary (Fig. 5-24). However, the amount of current flowing in the primary winding drops to a fraction of its load value, just enough to make up for transformer losses and supply a small magnetizing current. As far as the secondary is concerned, since it is open-circuited no current flows through it to a load. Nevertheless, the secondary voltage still appears across the winding. A similar situation exists in the case of every power outlet in the home. The full line voltage is available whether or not a load is connected to it. However, it is only when a load is plugged in that current flows. And, since power is the product of E and I, if there is no current in the secondary winding, there is no power.

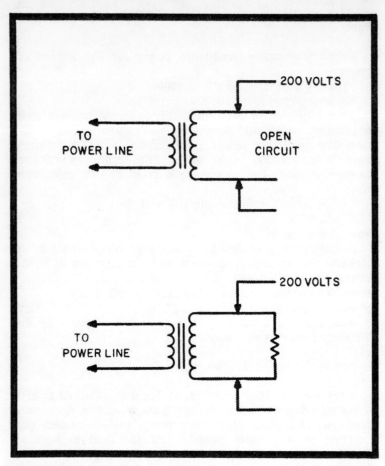

Fig. 5-24. The voltage across the secondary remains the same whether or not the secondary is connected to a load.

The Autotransformer

While the various transformer illustrations have shown four leads (two for the primary and two for the secondary), it is entirely possible for a transformer to have just three leads, or four leads, or a dozen or more. Some transformers have a single primary and a single secondary, but it isn't at all uncommon to have a transformer with one primary winding and four, five, or more secondary windings.

A transformer having the least number of leads is the autotransformer. As shown in Fig. 5-25, it consists of a single winding with a tap. The primary winding is from the bottom of the transformer to the tap (point B) while the secondary winding is from the bottom lead of the transformer to the top. This seems like a strange arrangement, but it works. Consider the primary connected to an alternator. A current will flow in the relatively few turns of the primary winding. The magnetic field accompanying this current will move or cut across the entire secondary winding and since the secondary has many more turns than the primary, there will be a voltage stepup, depending entirely on the turns ratio.

This means, of course, that the secondary current will also flow through the primary winding. This is not at all unusual in electronic circuits. Different currents frequently use the same conductor. Thus, it is entirely possible for a direct current and an alternating current to share the same wire and then for these currents to go their separate ways at some particular point in the circuit. If this were not possible, electronic circuits could become quite cumbersome.

The Multiple Transformer

A transformer can have a number of secondary windings. These can be all stepup or stepdown types, or any combination. Fig. 5-25 shows a transformer having one stepup winding and four stepdown windings. The primary winding may also be tapped. The purpose of the primary tap is to change the turns ratio. When tap 1 is used, the stepup turns ratio is increased. When tap 2 is used, the stepup turns ratio is decreased.

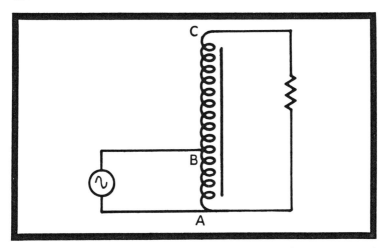

Fig. 5-25. The autotransformer uses a single winding for primary and secondary. The primary is the coil winding between A and B. The secondary is between A and C. Lead A is common to both windings.

165

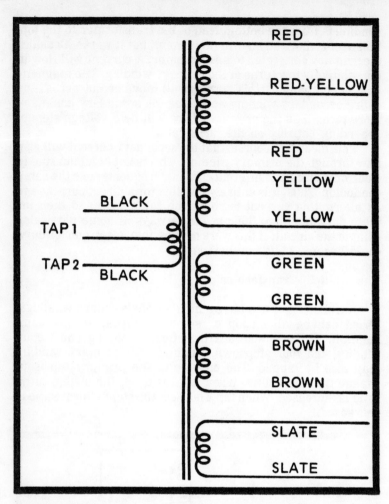

Fig. 5-26. Transformer with multiple secondary windings. The primary winding may be tapped for connection to different voltage sources. The red leads are high voltage, usually centertapped. Some of the other windings may also have a tap at the electrical center of the winding.

The wire connected to the center winding of a secondary is called a centertap. The centertap divides the secondary winding into two equal coils for use in certain electric and electronic circuits.

Transformer Color Code

Transformers are constructed according to the way in which they will be used in circuits. The leads of the transformers use wires covered with insulation carrying certain colors. Known as a transformer color code, it enables technicians to identify the various wires and connect them properly. Fig. 5-26 shows the color code for a power transformer.

PRODUCING VOLTAGES

The word voltage has a number of different names, some of which have already been used. Voltage is alternatively described as electrical pressure, electromotive force, voltage drop, IR drop, impressed voltage, voltage source, potential, and potential difference. But no matter what a voltage is called, there aren't too many ways of generating a potential. These potential generating techniques are 1) mechanical; 2) chemical; 3) heat; 4) friction; 5) light; 6) pressure of expansion. This doesn't mean these are the only techniques, for there are others, but the six methods mentioned here are the most common.

Mechanical Method

The alternator is a mechanical method of producing a voltage since it involves the mechanical motion of one of its parts. Basically, the generator consists of having the magnetic lines of flux of a magnet cut across the turns of wire of a coil. The coil can be made to move through the magnetic field or the magnetic field can be made to move across the coil. In either instance the physical movement of some part of the machine is involved.

Chemical Method

The battery is the prime example of producing an emf by chemical means. A battery consists of dissimilar substances placed in an electrolyte which can be either liquid or paste.

Heat Method

When two dissimilar metals are joined at one end (Fig. 5-27) and the joined end is heated, a voltage will appear across the opposite free ends of the metals. A device of this kind is called a thermocouple. Thermocouples are ordinarily used for temperature measuring and controlling devices. Although various metals are used in thermocouples, the only requirement is that the two metals be different. Thus, copper

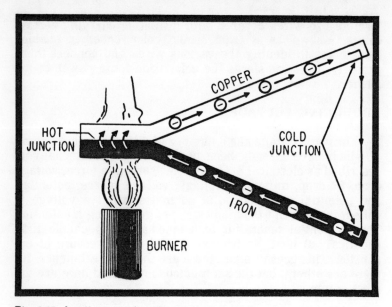

Fig. 5-27. A voltage can be generated when the junction of two dissimilar metals is heated. The voltage appears across the open ends of the metals. The arrows indicate the direction of current flow.

can be used as one of the metals, and iron for the other. The union of the two dissimilar metals is called a thermocouple junction.

Friction Method

It is almost impossible not to produce a voltage by this method since much of what we do involves friction. There is friction between the moving tires of an auto and the road. As a consequence the auto may develop a considerable potential, Discharge devices are often used to protect against such voltages. A comb moving through dry hair will produce enough voltage to create a visible spark or an audible crackling. Frictional electricity is also known as static electricity.

Light Method

Light energy can be changed into electrical energy by allowing it to shine on a photosensitive material. The photoelectric cell, or photocell, is one such device, and consists essentially of a metal disk acting as a support for a light-sensitive chemical. Various types of photocells are available:

one using copper oxide, another using selenium on iron. The copper oxide cell is made of copper covered with copper oxide. When light shines into the copper oxide it forces electrons from the oxide into the copper. Since the copper now has more electrons than previously it is negatively charged. Similarly, because the oxide has lost electrons, it is less negative than before, or saying the same thing, is positively charged. Because of this electron displacement, a voltage now exists between the oxide and the copper. The photocell, then, is a generator, with light supplying the energy necessary for moving the electrons from one region to another. All transistors are photoelectric cells if their internal junctions are exposed to light.

Expansion Method

Certain crystalline substances, such as quartz, tourmaline, or Rochelle salts can develop a voltage across the crystal faces if the crystal is subjected to some kind of pressure. This action, known as piezoelectricity, has a considerable number of practical uses. Crystals are used to produce ac voltages of constant frequency and so are used as the frequency determining elements of amateur and commerical radio stations. Crystals are also used in microphones where sound pressures cause a varying ac output proportional to the pressure applied. They are also used in phono pickups where they convert the mechanical movement of a stylus tracking the groove of a record into equivalent electrical voltages.

METERS

Meters and motors are included under one heading because both of these components operate on the same principle: the force of attraction and repulsion between magnets.

The basic meter known as the d'Arsonval, after its inventor, consists of a coil of wire wound on a bobbin suspended between the north and south poles of a horseshoe-type magnet. The basic construction details are shown in Fig. 5-28. The moving coil of the meter, sometimes called an armature, is mounted on a shaft and is free to rotate, but not in a complete circle. The movement of the coil is restricted by springs which are mounted on the shaft supporting the moving coil.

When a direct current is sent through the coil its effect is to make the coil into an electromagnet. The polarity of the

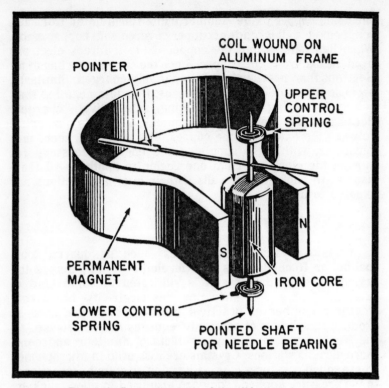

Fig. 5-28. Basic structure of the d'Arsonval meter.

magnet is such that its south pole is adjacent to the south pole of the enclosing horseshoe magnet. The north pole of the coil also faces the north pole of the permanent magnet. Since a force of repulsion now exists between the two magnets, they try to move away from each other. The permanent magnet cannot move since it is fastened into position and in any event would not move because of its weight. The movement of the coil is also restricted for it is designed to rotate in a clockwise direction.

Fig. 5-28 shows that a pointer is attached to the shaft on which the moving coil is mounted. Since the shaft will turn when the coil rotates, the meter pointer will move a certain distance. The amount of movement depends on the amount of current flowing through the moving coil. If the current is large enough, there will be a substantial deflection of the meter pointer. If the current through the armature coil is very small, the meter deflection may be barely noticeable.

When the current to the moving coil is interrupted—that is, when there is no current flowing through the coil—a pair of restraining springs return the coil to its original starting position. The meter pointer, of course, also moves back to its starting position.

All that is necessary now to complete the meter is to put a scale behind the pointer. A representative scale is shown in Fig. 5-29. The scale is then calibrated to indicate the current flowing through the meter coil. Thus 30 on the scale could correspond to 30 uA, 30 mA, or even 30A.

The d'Arsonval meter is a current reading meter or ammeter. It can be modified to measure voltage or resistance but the basic meter remains a current-reading type.

Types of Current-Reading Meters

Current-reading meters designed to read in amperes are called ammeters. By making a meter sensitive to smaller currents it can be made to respond to milliamperes, and is then called an milliammeter. A very sensitive current reading meter, the microammeter, is capable of measuring microamperes.

Meter sensitivity involves design and materials. The stronger the magnetic strength of the fixed-position horseshoe magnet and the greater the number of turns of wire of the armature coil, the more sensitive the meter. In such meters the clearance between the moving coil and the poles of the fixed magnet is usually very small. Further, the pivots of sensitive microammeters must be as friction-free as possible and so are often jewel mounted, much like the pivot arrangements in a quality watch.

Voltmeters

Just as an ammeter can indicate amounts of current, so too is it possible to have a meter measure voltage. Although

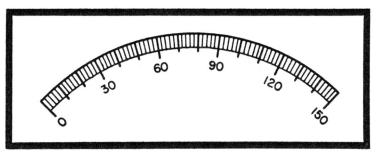

Fig. 5-29. One of the many types of meter scales that can be used.

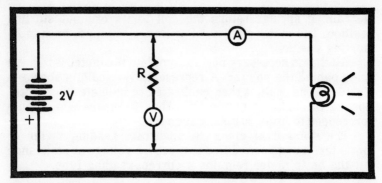

Fig. 5-30. The voltmeter (V), with the help of current-limiting resistor R (known as a multiplier resistor) measures the battery voltage. The ammeter (A) measures the current flowing from the battery through the lamp.

the instrument is called a voltmeter, it is simply the d'Arsonval current-reading meter adapted for this purpose. A resistor is put in series with the meter to limit the current. The scale behind the pointer is then calibrated in terms of volts.

Fig. 5-30 shows how a voltmeter and an ammeter are used in a simple circuit. The ammeter is connected in series with the 2V battery. The current flowing through the lamp also flows through the armature coil of the ammeter, and so the ammeter indicates the amount of current used by the lamp.

The voltmeter is connected in shunt or parallel with the voltage source. The resistor has a very high value, and so the current flowing through the armature coil of the voltmeter and resistor is very small. Thus, in the circuit shown in Fig. 5-30, the current through the ammeter and lamp might be ½A; the current through the voltmeter, though, could be less than 1 mA. The more sensitive the voltmeter, the smaller the amount of current it requires for deflection of its meter pointer.

MOTORS

Essentially, a motor is similar to a meter except that in the motor, the armature, corresponding to the moving coil of the meter, isn't restricted in its movement. The armature of the motor is allowed to rotate and since the armature is mounted on a shaft, the shaft can be coupled to some mechanical arrangement to perform some useful work.

With the meter, the problem of repulsion between the fixed permanent magnet and the electromagnet represented by the armature is easily solved. With the motor, though, the rotating armature could easily turn 180 degrees and present its north pole to the south pole of the fixed magnet. At this point rotation would stop since the opposite poles of the two magnets would attract each other.

Fig. 5-31 is a stripped down drawing of a motor with the numerous turns of the armature coil represented by a single loop. The armature coil is lettered A-B-C-D and the direction of rotation is shown by the curved arrow. The armature receives its current from a battery and, since the battery is fixed in position while the armature rotates, there must be some way of connecting the battery to the armature. This is done by a commutator consisting of two segments as shown. Resting against the commutator are a pair of carbon brushes which make electrical contact but do not restrict the motion of the commutator.

With the current flowing through the armature coil, the north and south poles of the coil are as shown in the illustration. The north pole of the armature is being repelled by the north pole of the permanent magnet (sometimes called a field magnet). Similarly, the two south poles are facing each other; so there is a force of repulsion here. As a result the armature will try to move out of the way and will do so by turning in a clockwise direction.

The coil will swing around and make a half turn. Theoretically, the south pole of the armature coil should now face the north pole of the field coil. However, the split-ring commutator transposes the connections to the voltage source

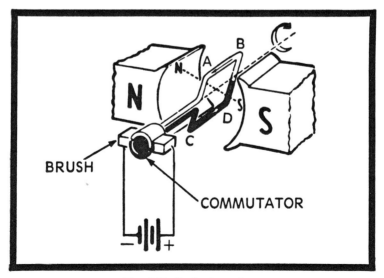

Fig. 5-31. Basic dc motor.

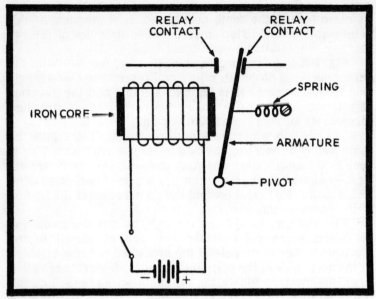

Fig. 5-32. Basic elements of a relay. When the switch is closed, current flows through the relay coil and attracts the armature.

and so the current through the armature coil changes its direction.

In an actual motor, a large number of individual coils are used, with each coil having its own pair of commutator segments. As a result, there are always a few coil segments facing the field magnet with the result that there is constantly a force of repulsion between the armature and the field magnet.

RELAYS

The force of attraction (or repulsion) between magnetic poles is a simple enough idea, but it has resulted in tremendous advances in electronic devices. Meters and motors are just two examples of the important applications of this basic magnetic principle. Relays represent still another use.

In its basic elementary form, a relay (Fig. 5-32) consists of a coil, a moving metallic element called an armature, a spring, and one or more contacts. Unlike the armature of a meter or motor, the armature of a relay is pivoted at one end with the opposite end free to move up and down or from one side to the other. The relay coil is connected to a voltage source and so when a current flows through the relay coil it becomes an electromagnet. At this moment no current flows through the relay coil and so the armature is held against the right contact by the force of the spring. When a current is allowed to pass through the coil it becomes an electromagnet and attracts the armature. The strength of magnetic pull is great enough to draw the armature against the force of the spring. The armature now touches the left contact connection and will remain in this position as long as current continues to flow through the coil. When the switch is opened, the electromagnet will become an ordinary nonmagnetic coil again and the armature will resume its original position with the spring pulling the armature back to its starting contact.

Fig. 5-33 shows how a relay can be used to open and close two different circuits. While this could also be accomplished manually by a switch, or a pair of switches, it is often convenient and practical to be able to do so by using a relay. The relay, for example, might be mounted in an area where chemical fumes or temperature would make manual operation hazardous. Relays can also be controlled remotely by radio signals and so they have found extensive applications in aircraft and space vehicles.

Like other electric and electronic components, there are a tremendous variety of relays, some of which are designed to

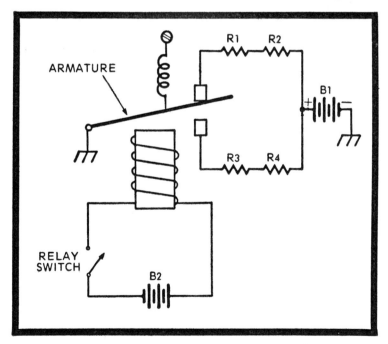

Fig. 5-33. A relay can be used to switch a pair of circuits on or off, one at a time, as in this diagram.

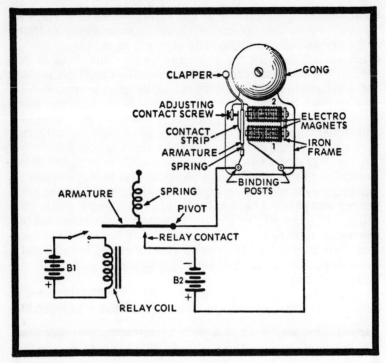

Fig. 5-34. The bell rings when switch is closed.

open or close a large number of circuits, or to hold a circuit open or closed for a predetermined amount of time. A relay, then, is an electromechanical switch and is related to the meter and motor since, like these other devices, it makes use of the magnetic forces of attraction or repulsion.

In Fig. 5-33 resistors R1 and R2 represent one circuit and resistors R3 and R4 another circuit. When the relay switch is open, no current flows through the relay coil and the armature of the relay touches the upper contact, closing the R1-R2 circuit. Current now flows from the negative terminal of battery B1 to a common connection. One end of the metallic armature is also grounded and so current flows through the armature to the upper contact of the relay, and then through R1 and R2. This current will continue flowing as long as the relay switch is open.

When the relay switch is closed, current flows through the relay coil, making it an electromagnet. The magnet attracts the armature, thus disconnecting the upper relay contact and closing the lower one. Current now flows from battery B1 to ground, over to the armature, through the armature, and then the lower relay contact, continuing through R3 and R4. In this way, current can be controlled—with current going through R1-R2 or R3-R4 as desired.

In another application, a relay with a single contact is used to control the ringing of a bell. In Fig. 5-34 current for the bell-ringing mechanism is supplied by battery B2. When the spst switch in the relay circuit is closed, the armature pulls away from its open position, and touches the lower contact. This completes the bell circuit.

Since relays are electromechanical switches, many of them are described in the same way that switches are identified. In Fig. 5-35A, for example, the relay is a single-pole, single-throw type (spst). This relay is a normally closed type (n.c.). With no current flowing through the relay coil, the relay contacts touch each other. It is only when current is allowed to flow through the relay coil that the contacts are separated. Drawing B shows the same type of relay, but a normally open (n.o.) type. C shows a single-pole, double-throw type and D a double-pole, single-throw type.

Relay arrangements can become complex and drawing E is an example of a relay designed for opening and closing a large number of circuits. This is a triple-pole, double-throw unit (tpdt).

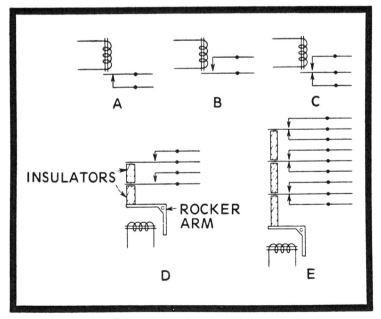

Fig. 5-35. Types of relays and their symbols.

The study of electricity often proceeds in a circular fashion. In an earlier chapter batteries were described as electrical storage devices. But if you'll recall the Leyden jar mentioned previously, you'll remember that there is another storage device, the capacitor. Although capacitors are basically very simple units, they are as important as resistors and coils and warrant a chapter of their own. We call it Chapter 6.

How Electricity Is Stored

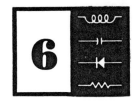

When Ewald George von Kleist, at one time dean of the cathedral of Kamin in Pomerania and Professor Pieter van Musschenbroek of the University of Leyden, in Holland, independently invented the Leyden jar, neither realized the far-reaching effects of their discovery, for what they had actually done was to produce the first capacitor. Electrical and electronic apparatus consists of relatively few parts (but great varieties of them): tubes, solid-state components, resistors, coils, insulators and conductors, and capacitors. It is doubtful if electricity and electronics would have made much progress without the capacitor, just as it is difficult to imagine a kitchen without glasses, jars, jugs, pots and pans. A capacitor is a device that can store electrons, somewhat analagous to a glass or bottle holding water.

The Leyden jar, described in Chapter 2, was the first true capacitor. However, one investigator, Dr. John Bevis, soon realized that the shape of the jar had nothing to do with the way it behaved and so he replaced the jar with a flat section of glass and put sheets of metallic foil on both sides. You will find variations of this extremely simple device associated in some way with a tremendous number of electrical and electronic devices. No radio, television set, or computer is made without them. Every automobile and plane uses them.

THE BASIC CAPACITOR

Basically, then, a capacitor consists of a nonconductor, or insulator, placed between two conductors, generally metal plates. (Fig. 6-1.) The insulating material is called a dielectric, with just about every different kind being tried for the job. Dielectrics used in modern capacitors include paper, mica, ceramic materials such as titanium dioxide and barium titanate, Mylar coated paper, thin plastic film or thin oxide film, Bakelite, castor oil, mineral oil, and air. For the capacitor plates, the metals can be stamped aluminum, smooth aluminum foil, etched aluminum foil and silver.

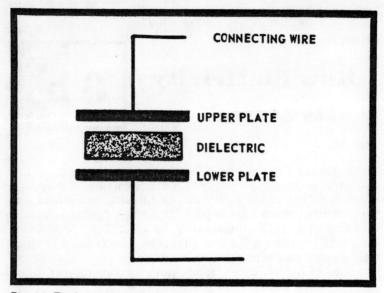

Fig. 6-1. The basic capacitor consists of a pair of metal plates separated by an insulating material, the dielectric.

CHARGING THE CAPACITOR

A battery is a chemically operated device for storing a tremendous number of electrons on one electrode obtained by stripping an adjacent electrode of its electron supply. Because of this electron imbalance a voltage exists between the two electrodes. The electron-rich electrode is minus or negative; the electrode with an electron shortage is plus or positive. The

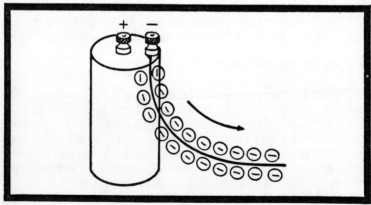

Fig. 6-2. Electrons motion is from the negative terminal of the cell through the wire.

electrons, crowded on the negative electrode will, if given the opportunity, either make an attempt to return to their original home, the positive electrode, or if that isn't possible, will spread out on any available metallic surface.

If you connect a copper wire to the negative terminal of a cell, electrons from that terminal will move along the wire from its start to finish. (Fig. 6-2.) The arrow indicates the initial movement of this electron current. The flow of electrons stops rather quickly, since the amount of room for electron movement is so limited. Those electrons that spread out on the loose wire oppose the further migration of electrons from the negative terminal of the cell.

If a metal plate is attached to the loose end of the wire, there will be another momentary surge of electrons since more surface area has been provided. And, as in the case of the wire, electrons will move into the metal and onto its surface. Once again, an electrical back pressure will build up, preventing further transfer of electrons to the metal plate. This electrical back pressure could be called a counter emf or an opposition voltage.

Fig. 6-3 shows just one plate of a capacitor. There is also a dielectric, although it is invisible. The dielectric in this case is air, but the air could be replaced by glass, rubber, mica, etc.

THE DIELECTRIC

Like all matter, the dielectric is composed of molecules which, in turn, are made up of atoms. Each atom has a central

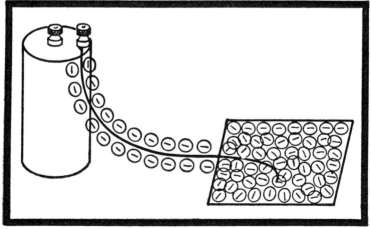

Fig. 6-3. If a metal plate is attached to the wire, there will be a further momentary motion of electrons crowding onto the metal plate.

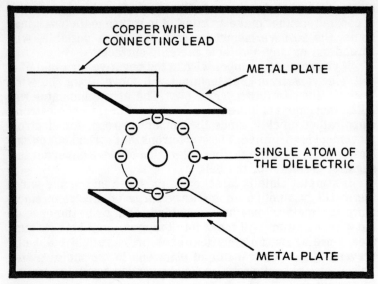

Fig. 6-4. Each atom of the dielectric has a central nucleus surrounded by electrons.

portion or nucleus and various rings of orbital electrons. To simplify matters Fig. 6-4 shows a single atom with its outermost electron ring. The assumption is made here that the electrons are moving in a circle around the nucleus. At this moment the two metal plates are not connected to a cell.

ACTION OF THE CAPACITOR

With the capacitor now connected as in Fig. 6-5 the electrons flow from the negative terminal of the battery, through the connecting wire, occupying both the wire and the plate. The electrons in orbit around the atom in the dielectric no longer follow the same orbital path, but are pushed away as far as possible from the top plate. These orbital electrons now come closer to the lower plate. Since electrons repel each other, electrons that originally occupied the lower plate now move away from it and migrate toward the positive terminal of the battery. The effect of the charged upper plate, then, is to force electrons to leave the lower plate.

This circuit has two ammeters, marked A, instruments for measuring the amount of current flow. At the moment the capacitor is connected to the battery, the pointers of both meters flick upward to the same extent, and then move back to zero. This indicates the same amount of current flows in the upper wire as in the lower one; the amount of electron current flowing into the upper plate of the capacitor is equal to the amount of electron current flowing out of the lower plate.

The fact that both meter pointers drop to zero indicates that after this transfer of electrons, all electron movement stops. At this time the voltage measured from the upper capacitor plate to the lower one is exactly the same as the battery voltage and has the same polarity. The capacitor is now fully charged, and this charge or voltage opposes any further electron movement.

If the capacitor is disconnected from the battery, the stored electric charge is portable. The disconnected capacitor can be attached to some circuit and will then deliver its electrons to it, becoming discharged in the process.

The capacitor can be discharged rapidly by short circuiting it—that is, by connecting a conductor, such as a wire, from one plate to the other. The capacitor can also be discharged somewhat more slowly by connecting a resistor from one plate to the other. The higher the value of resistance, in ohms, the longer it will take for the capacitor to reach zero charge. Zero charge, or zero voltage exists when the number of electrons on both plates is approximately the same.

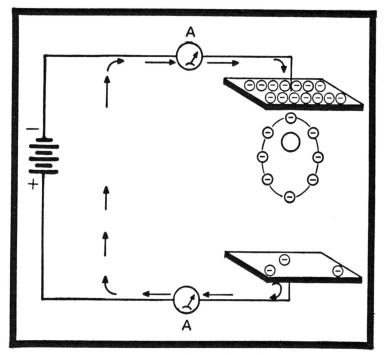

Fig. 6-5. Here electrons are forced to leave the lower plate.

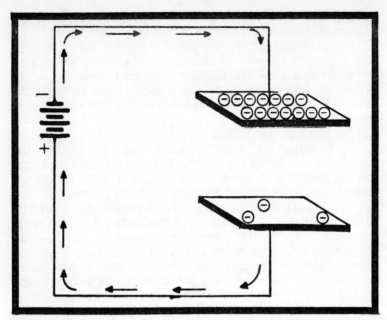

Fig. 6-6. Both plates have electrons, but since the upper plate has more, it is negative with respect to the lower plate. The lower plate is plus; the upper, minus.

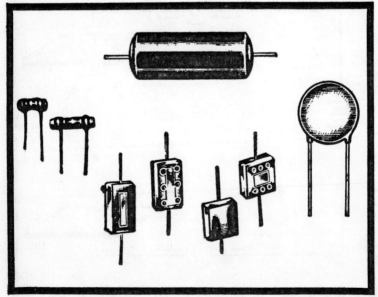

Fig. 6-7. These are just a few of the many different kinds of fixed capacitors.

When a capacitor charges, its voltage increases from zero (when the capacitor is first connected) to the voltage of the source. When a capacitor is discharged, the capacitor voltage decreases to zero. The amount of voltage of the capacitor (Fig. 6-6) depends on the electron difference between the two plates. The greater this difference, the greater the voltage.

FIXED AND VARIABLE CAPACITORS

A fixed capacitor, as its name implies, is one whose plates and dielectric cannot be moved. A variable capacitor is a unit having two plates or two sets of plates. The usual arrangement is to have one set of plates interleave the other. The "stator" plates remain fixed in position while the other set, the rotor, is mounted on a shaft which can be turned. Fig. 6-7 illustrates different types of fixed capacitors while Fig. 6-8 shows variable units. In some variable capacitors there is just a single rotor and a single stator, but two or more variable capacitors can be operated by a single shaft. In this way a number of sets of rotor plates can be turned simultaneously, with all stator plates remaining fixed, attached to a frame which holds them securely in position. The shaft is operated by a knob or a dial in front of the electrical or electronic equipment. As the knob is turned, the capacitance is changed and is minimum when the rotor plates are completely unmeshed from the stator plates and is maximum when the plates are meshed.

Still another type of variable capacitor resembles both fixed and variable capacitors. Known as a trimmer or padder, depending on its construction and the work it must do, the unit is adjusted by a special type of tool made of a nonconducting material. The capacitor is varied by a setscrew. This type of capacitor uses either air or mica as the dielectric and may have two or more plates. There are various other capacitors of this kind but those in Fig. 6-8 illustrate the general principle.

Capacitor Symbols

Fig. 6-9 shows the various symbols for fixed and variable capacitors. One capacitor type, the electrolytic, to be described later, generally uses a plus symbol, while the others do not.

The most commonly used symbols for fixed capacitors are those shown as 1 and 2 in Fig. 6-9A. These symbols are modified, as indicated in drawing 3, in circuit diagrams intended for industrial or commercial use.

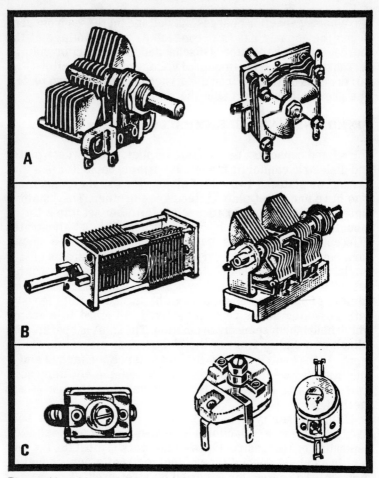

Fig. 6-8. Variable capacitors can be single units (A) or ganged (B). Some variables use a screw-type adjustment (C).

The symbol identified by the number 4 is a dual or double capacitor having one common plate with the dielectric between it and two other plates. The capacitors represented in 1, 2, and 3 each have a pair of connecting leads, but the capacitor in symbol 4 has three leads. One lead, often made of wire covered with black insulation, is called the common, while the other two may be referred to as "hot" leads. This type of wire coding isn't always used, but generally the two "hot" leads have the same color to distinguish them from the common lead.

The symbol in 5 is a "feedthrough" capacitor. This capacitor often has a ceramic material as its dielectric and

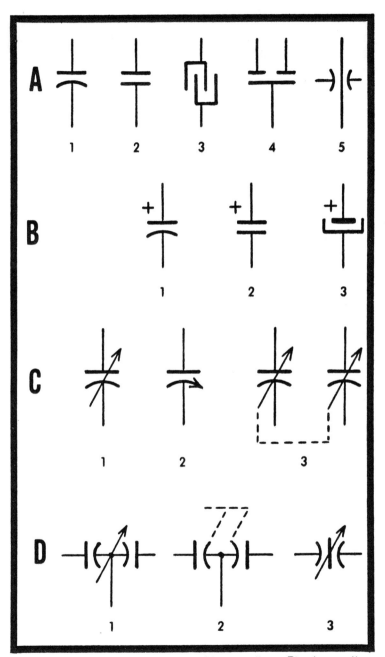

Fig. 6-9. Symbols for fixed and variable capacitors. Fixed capacitors (A); electrolytics (B); variable capacitors for receivers (C); variable capacitors for transmitters (D).

can thus supply a relatively large amount of capacitance in a small space. The capacitors are called feedthrough since they are designed for single-hole mounting.

The three capacitors in row B are all electrolytics. The fact that the capacitor is an electrolytic type is indicated by the plus sign placed above the plate symbol. Except for the plus sign, the two symbols, 1 and 2, are exactly the same as the first two symbols shown in row A. However, symbol 3 in row B is specifically used for electrolytic capacitors only. The short horizontal straight line always represents the plus terminal of the capacitor and so the plus symbol is sometimes omitted.

Rows C and D in Fig. 6-9 illustrate the symbols used for variable capacitors. Symbols 1 and 2 in row C are for single variable capacitors, that is, capacitors having but a single stator and rotor.

The symbol for two "ganged" capacitors (two on one shaft) is represented by symbol 3, row C. The dashed line is used to indicate that the rotor shaft is common to the two variable capacitors. If three variable capacitors have a single shaft the symbol is exactly the same as that shown in drawing 3, row C, except that another capacitor symbol is added and the broken line is extended to include it.

Quite commonly, in a circuit using a two- or three-ganged capacitor, the capacitors may be shown widely separated. However, the fact that the capacitors are actually physically adjacent and use the same rotor shaft is always indicated by a broken line. The variable capacitor symbols shown in row C are used for radio receiver circuit diagrams, while the variable capacitor symbols in row D are ordinarily found in circuit diagrams for transmitters.

CAPACITOR CODES

The letter C is often used in conjunction with capacitor symbols in circuit diagrams. When a number of capacitors are used, they can be identified as C1, C2, C3, etc. It is usual practice to start with capacitor C1 at the upper left-hand side of a circuit, but this isn't a practice always followed.

UNITS OF CAPACITANCE

Capacitance refers to the amount of charge a capacitor can store. The basic unit of capacitance is the farad (F), named in honor of Michael Faraday (1791 - 1867). The farad, however, is such an extremely large unit that submultiples are usually required. A more practical value of capacitance is the microfarad (uF), equal in value to one-millionth of a farad. Sometimes even this submultiple is unwieldy, so we have nanofarads (nF) and picofarads (pF). A nanofarad is one-thousandth of a microfarad; a picofarad is one-millionth of a microfarad. Formulas in electricity and electronics, though, are often expressed in terms of farads, and so it is sometimes necessary to be able to convert to and from farads.

To change from farads to microfarads, multiply farads by one million. And to change from microfarads to farads, divide microfarads by one million. This involves moving the decimal point to the left or to the right six decimal places. The important factor in converting from farads to microfarads or microfarads to farads is to remember in which direction to move the decimal point and to move it the correct number of places.

FACTORS THAT DETERMINE CAPACITANCE

The amount of capacitance is determined by the area of the plates in the capacitor, by the kind of dielectric used, and the distance the plates are from each other. The greater the plate area, the larger the amount of capacitance. Large plate areas can be obtained by using a number of plates and connecting them. Thus, the top plate of a capacitor, for example, may not be just one plate, but a number of them in parallel with each other, all connected by a conductor. The bottom plate is made the same way. The plates are interleaved but are separated by a dielectric material, such as paper. Another technique for increasing capacitance is to use a very thin material, such as aluminum foil, in roll form (Fig. 6-10). The foil-paper-foil combination is tightly rolled. The two very long lengths of foil do not touch each other but are separated by the paper. Even more capacitance is obtained by chemically etching the aluminum foil surface to increase its area.

Fig. 6-10. A capacitor can be made by using two strips of foil separated by paper. The combination is rolled, put into a tubular paper container which is then waterproofed.

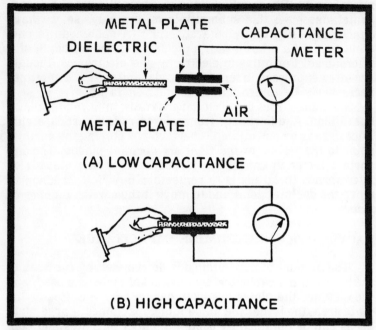

Fig. 6-11. The amount of capacitance may be determined by the dielectric. The dielectric in A is air. The capacitance is increased when a different dielectric material, such as mica, is inserted between the capacitor plates.

Another way of increasing capacitance is to bring the plate materials as close to each other as possible; the closer, the greater the capacitance.

Finally, the capacitance is influenced by the kind of material used as the dielectric. Air has a dielectric value of 1. Mica has a dielectric value of 5 to 10; so, if mica is substituted for air as the dielectric, the amount of capacitance will be multiplied anywhere between 5 and 10 times. The ability of a dielectric to increase capacitance is called the dielectric constant and is represented by the lowercase letter k. Air has a k of 1; mica ranges anywhere between 2 and about 8 or 10 depending on the kind that is used; ceramic materials have a k that extends from 80 to 1200; paper has a k of about 2. 1 uF capacitor using air as a dielectric will have a value of 2 uF if paper is substituted for air (Fig. 6-11).

Electrolytic Capacitors

Capacitors can all be grouped under the two general headings of fixed and variable. And under the heading of fixed capacitors we can once again have two main groups: the electrostatic types (those having Mylar, paper, mica, ceramic, and the like as the dielectric), and the other electrolytic.

In the electrolytic capacitor, an electrochemical process is used to form a microscopic film of oxide on one of the metal plates. This oxide coating serves as the dielectric and is sandwiched between two metallic plates, often long lengths of aluminum foil, compactly rolled up before being put into its case.

Electrolytic capacitors are characterized by very high capacitance compared to most electrostatic types. They are also polarized, and must be properly connected into circuits with the plus terminal of the electrolytic capacitor connected to the plus terminal of the dc voltage source and the minus terminal attached to the minus terminal of the voltage. Electrolytic capacitors and electrostatic types are not always interchangeable, but when they are, the electrolytic must be substituted with consideration of voltage polarities.

TANTALUM CAPACITORS

Tantalum capacitors are members of the electrolytic family. In these capacitors a tantalum anode (Fig. 6-12) serves as the positive plate of the capacitor; an oxide film, electrochemically formed on the tantalum surface, is the dielectric; and a deposited layer of semiconducting manganese dioxide forms the cathode plate of the capacitor. Conductive coatings are bonded to the manganese dioxide to allow an electrical connection from the cathode to the case. The chief advantage of this capacitor is its stability and ratio of size to rating. Large capacitance values are available in extremely small packages.

WORKING VOLTAGE

The ability of a capacitor to withstand the voltage put across its terminals is known as dc working voltage and is abbreviated as DCWV. A capacitor that is marked 450 DCWV indicates that the capacitor may safely be connected across a dc voltage source having a maximum output of 450V. The working voltage depends on the way the capacitor is manufactured, the type and thickness of dielectric and on the temperature.

Two opposing factors are at work here. The thinner the dielectric, the greater the capacitance. But the thinner the

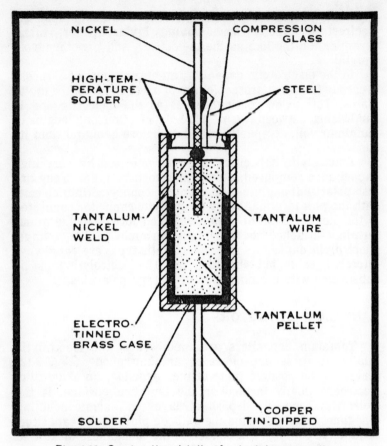

Fig. 6-12. Construction details of a tantalum capacitor.

dielectric, the greater the opportunity for the impressed voltage to puncture the dielectric. This isn't the entire answer, though. Some dielectrics can tolerate higher voltages than others (called "dielectric strength"). A ceramic material, for example, has a higher dielectric strength than most kinds of mica. Air is about the weakest kind of dielectric while flint glass is one of the strongest. However, the dielectric constant must also be taken into consideration, for while a dielectric may tolerate a high voltage, its value of k may also be very low.

The dc working voltage of a capacitor is usually specified within a certain operating temperature range. Ordinarily, if the temperature at which the capacitor is to work is much higher than the specified dc working voltage temperature, the working voltage is reduced.

When a capacitor is to be used in a circuit containing ac, the maximum or peak voltage is also specified by the manufacturer of the capacitor.

COLOR CODES FOR CAPACITORS

As in the case of resistors, capacitors can be color coded but unfortunately there are a variety of codes. The color code for commercially coded capacitors is the same as that for resistors. Capacitors for military use may be coded following military specifications, or a manufacturer may use a code of his own for particular purposes. In some instances, especially if the capacitor is extremely small, as it would be in subminiature circuits, no coding is used at all. Large capacitors,

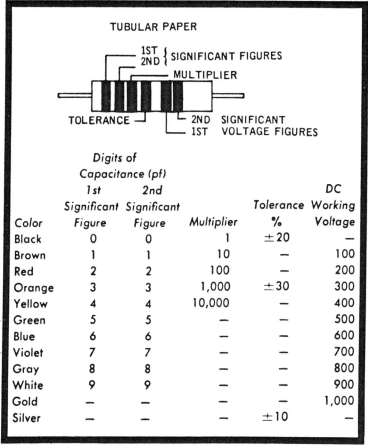

Fig. 6-13. Color code for molded paper tubular capacitors.

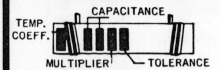

5 - DOT RADIAL LEAD CERAMIC

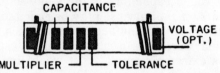

BYPASS OR COUPLING CERAMIC

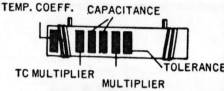

SIX - DOT CERAMIC

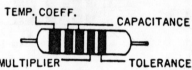

Tubular Ceramic Capacitor Five-Color System

First color	Temperature Coefficient of Capacitance
Second color	First significanct figure of capacitance
Third color	Second significant figure of capacitance
Fourth color	Decimal multiplier of capacitance
Fifth color	Tolerance of capacitance

Tubular Ceramic Capacitor Six-Color System

First color	Significant figure of temperature coefficient of capacitance
Second color	Multiplier to apply to significant figure of temperature coefficient
Third color	First significant figure of capacitance
Fourth color	Second significant figure of capacitance
Fifth color	Decimal multiplier of capacitance
Sixth color	Tolerance of capacitance

Code for Ceramic Dielectric Capacitors (continued)

Color	Temperature Coefficient of Capacitance (5 Dot System)	Significant Figure of Temperature Coefficient of Capacitance (6 Dot System)	Multiplier to Apply to Significant Figure of Temperature Coefficient (6 Dot System)
Black	0 pts/mln/°C	0.0	−1
Brown	− 33 pts/mln/°C	—	−10
Red	− 75 pts/mln/°C	1.0	−100
Orange	−150 pts/mln/°C	1.5	−1000
Yellow	−220 pts/mln/°C	2.2	−10000
Green	−330 pts/mln/°C	3.3	+1
Blue	−470 pts/mln/°C	4.7	+10
Violet	−750 pts/mln/°C	7.5	+100
Gray	General Purpose	General Purpose	+1000
White	General Purpose		+10000

Color	1st and 2nd Significant Figure of Capacitance	Decimal Multiplier of Capacitance	Tolerance of Capacitance Nominal 10 pf or Less	Tolerance of Capacitance Nominal Over 10 pf
Black	0	1	±2.0 pf	±20%
Brown	1	10	±0.1 pf	±1%
Red	2	100		±2%
Orange	3	1000		±3%
Yellow	4	10000		
Green	5		±0.5 pf	±5%
Blue	6			
Violet	7			
Gray	8	0.01	±0.25 pf	
White	9	0.1	±1.0 pf	±10%

Fig. 6-14. Color code for five- and six-color tubular ceramic capacitors.

such as electrolytics, may have data concerning the capacitor printed or stamped directly onto the capacitor case. In some instances, capacitors and resistors look so much alike that it is sometimes difficult to identify them, particularly if they aren't coded.

A tubular paper capacitor might use rings of color to supply information. These capacitors (Fig. 6-13) can have five or six color rings. When reading the capacitor color code, the capacitor must be held so the color rings are at the left.

If just one color ring is used at the extreme right—that is, following all the other color rings—the color ring indicates the dc working voltage, as shown in the column at the right in the table. Thus, if the last color is blue, the dc working voltage is 600V. However, if two colors are used instead of just one, consider that they represent the first two significant digits of the voltage and follow them with two zeros. Thus, if the last two color rings are brown and red (brown = 1, red = 2), the dc working voltage (DCWV) is 1,200 volts.

Tubular ceramic capacitors use color rings, color rectangles, or color dots (Fig. 6-14). Either five or six colors are used.

The body of a ceramic disc or Mylar capacitor is often colored to indicate its working voltage. Green indicates low-voltage (100V or less), and most transistor radios use capacitors of this color. Orange indicates a higher voltage (up to 600V). But this rule is not an industry standard and variations are not uncommon.

CAPACITOR LEAKAGE

Since by their very nature capacitors are constantly subjected to electrical pressure, it isn't too surprising to find that electrons sometimes find their way from the charged plate through the dielectric to the uncharged plate. Electrons can also move through any conducting substance accumulated around the capacitor leads, such as dust containing metallic particles. Tiny impurities in the dielectric can also result in a current movement from plate to plate inside the capacitor. The sum total of all these electron migrations is called leakage. Whether the leakage is a serious problem or not depends on the amount of leakage and the work the capacitor is expected to do. Electrolytic capacitors have more leakage than electrostatic types.

Leakage is minimized by using thicker dielectrics, or dielectrics that have greater insulating properties, or by sealing the capacitor to make it impervious to moisture.

INCREASING CAPACITANCE

Increasing or decreasing capacitance with a variable capacitor is no problem. Maximum capacitance is obtained when the capacitor plates are fully meshed; minimum capacitance when the plates are unmeshed. Note that when the plates are unmeshed the capacitance does not drop to zero, but to some small value compared to full capacitance.

For fixed capacitors, the only way to increase capacitance is to wire two or more capacitors in parallel with each other, a method also known as a shunt connection. The total capacitance is then equal to the sum of the individual capacitances. If a 4 uF capacitor is wired in parallel with a 6 uF (Fig. 6-15) the total capacitance is 4 + 6, or 10 uF. However, both capacitors should have the same DCWV rating. For particular applications, an electrostatic capacitor is sometimes shunted across an electrolytic type. Electrolytics are frequently paralleled to obtain larger amounts of capacitance. Capacitors to be wired in shunt can have identical values or may be completely different.

Although Fig. 6-15 shows just two capacitors in parallel, any number may be wired this way, with the total capacitance obtained by adding the values of all the individual capacitors. As a practical matter, it is rather unusual to find more than two capacitors placed in shunt.

When adding the capacitance values to get the total capacitance, it may be necessary to convert, if different

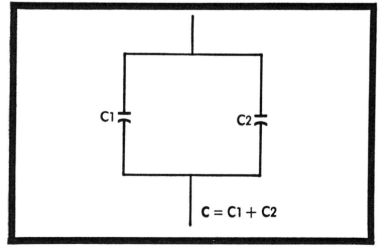

Fig. 6-15. The total capacitance (C) of capacitors wired in parallel is equal to the sum of the individual capacitances.

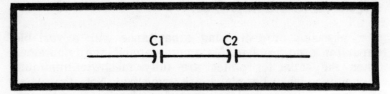

Fig. 6-16. Series capacitor arrangement.

capacitance multiples are used. Thus, if a 2 uF capacitor is in shunt with a 3600 pF unit, either convert to microfarads or to picofarads before making the addition.

You will note that the method for finding the total value of parallel-connected capacitors is the same as the method for determining series-connected resistors—straight addition. As might be expected, the method for calculating the value of series-connected capacitors is the same as that used for parallel-connected resistors. Let's examine this in more detail.

DECREASING CAPACITANCE

To decrease overall capacitance, capacitors can be wired in series, as indicated in Fig. 6-16. The total capacitance is then smaller than that of the unit in the series group having the least capacitance. And to find the total capacitance (C) of two capacitors in series we use the same general formula described earlier for two resistors in parallel.

$$C = \frac{C1 \times C2}{C1 + C2}$$

Step 1: Multiply the capacitance values of the two capacitors.
Step 2: Add the capacitance values of the two capacitors.
Step 3: Divide step 1 by step 2—that is, divide the product (the result of the multiplication) by the sum (the result of the addition).

The only requirement is that the two capacitors use identical capacitance units. Both capacitors must be in pF, uF, or in farads. If one of the capacitors is in pF and the other in uF, either convert pF to uF or uF to pF before doing the arithmetic.

Sometimes three or more capacitors are wired in series. To find the total capacitance, work with two capacitors at a time. Find the total capacitance, and then, regarding this total capacitance as a single unit, combine it with one of the remaining capacitors in the series string.

Often, a technician working on a transmitter or receiver circuit will make his own capacitor (which he calls a "gimmick") by connecting two short pieces of insulated wire to the terminals where the capacitance is required. He can make the wires as long as he likes to get as much capacitance as required—usually no more than a few picofarads. He twists the wires tightly together so their position relative to each other will be stable, then cuts the twisted pair's length until the capacitance value is what he wants.

STRAY CAPACITANCE

A capacitor is formed whenever two metal surfaces are near each other. Thus, in electrical or electronic apparatus, a bare wire adjacent to a transformer case, a capacitor case, or a metal chassis, represents a certain amount of capacitance, with air as the dielectric. If a pair of wires are near each other, and the wires are insulated with cloth or rubber, the insulation material becomes the dielectric. The capacitance, known as stray capacitance, does not affect dc or low-frequency alternating currents, but can change the operation of high-frequency circuits. Stray capacitance becomes more important as frequency becomes higher. In very high frequency circuits, designers must be very careful about the positioning of wires that carry current. The general rule is to make these wires as short as possible.

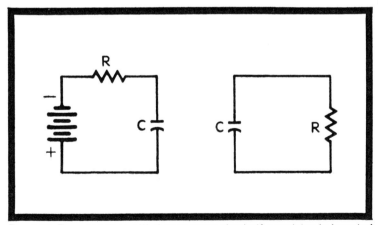

Fig. 6-17. The capacitor will charge more slowly if a resistor is inserted between it and the voltage source. The larger the value of resistance, the longer it takes the capacitor to charge. A charged capacitor can also discharge through a resistor. The larger the resistance value, the longer it takes for the capacitor to discharge.

CONTROLLING CHARGE AND DISCHARGE

If a resistor is inserted between a capacitor and battery, the capacitor will charge more slowly (Fig. 6-17). To find out how long it will take (in seconds) for the capacitor to charge to 63 percent of its maximum value, multiply the value of resistance (in ohms) by the value of capacitance (in farads).

Example:

How long will it take, in seconds, for a 2 uF capacitor to charge if the resistor in series with it is 100K? 2 uF = 0.000002 farad. So time (in seconds) is 0.000002 x 100,000 = 0.2 second.

Thus, this capacitor will be charged to 63 percent of its maximum value in 200 milliseconds. The time it takes for a capacitor to be charged to 63 percent of its maximum charge is called a time constant. Note that the time it takes to charge a capacitor isn't influenced by the voltage of the source. The 2 uF capacitor in the example given above will require 0.2 second to charge to 63 percent of its maximum charge whether it is connected to a 1.5V dry cell or a 90V battery.

The reason 63 percent is used is that a capacitor is considered charged when it stores 63 percent of the charging voltage. This is a practical figure. Naturally, if it takes 200 milliseconds (msec) to charge the 2 uF capacitor to 63 percent of maximum, the capacitor will ultimately be charged to the fullest amount if it is allowed to remain connected to the voltage source for a much longer time than this.

When a capacitor is fully charged, the rate at which it will discharge depends on the amount of capacitance and the value of resistance placed across the capacitor. A capacitor will discharge down to 37 percent of its full charge in one time constant.

Example:

How long will it take a capacitor rated at 5 uF to discharge to 37 percent of its full charge if it is shunted by a 50K resistor? 5 uF is 0.000005 farad; and time (in seconds) is 0.000005, so the time is 0.25 sec, or 250 msec.

CAPACITORS IN DC AND AC CIRCUITS

The behavior of a capacitor in an ac circuit is quite different than the way it acts in a dc circuit. The arrangement of Fig. 6-18 shows a lamp in series with a resistor, with both

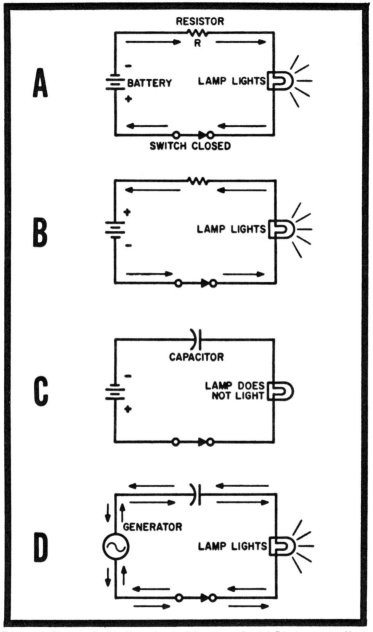

Fig. 6-18. Reversal of battery leads (drawings A and B) does not affect the flow of current, just its direction. The capacitor blocks the passage of current in the dc circuit of C, but in the ac circuit of D the current moves back and forth.

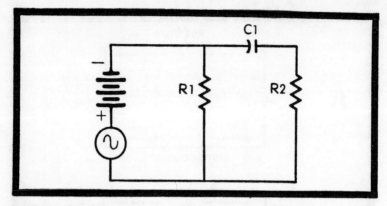

Fig. 6-19. Capacitors can control the flow of current. Two currents, one ac and one dc, will flow through R1 since it is across both voltage sources. Capacitor blocks dc to R2, but couples ac through R2. In this application C1 is called a blocking capacitor for dc and a coupling capacitor for ac.

connected to a battery. There is no problem about current flowing and it moves in the direction indicated by the arrows. The lamp lights, and remains lighted, as long as current flows through it.

In drawing B the battery is transposed and so current moves in the opposite direction. This does not affect the operation of the lamp and again, it remains lighted.

When a capacitor is substituted for the resistor in drawing C, there is a momentary surge of current. This charges the capacitor but the voltage across the capacitor becomes equal and opposite that of the battery voltage. In this circuit, then, there are two factors which stop the flow of current. One of these is the charge on the capacitor and the other is the capacitor dielectric. Since this is a nonconductor, current cannot flow through it. The lamp in drawing C does not light.

When an ac generator is substituted for the battery, as in drawing D, the lamp lights. This does not mean that current flows through the capacitor. Since the generator reverses its polarity regularly, the current surges back and forth in the circuit, flowing through the lamp each time it does so. The capacitor charges and discharges regularly. Thus, a capacitor in an ac circuit does not prevent the flow of current while it does so in a dc circuit.

SEPARATING AC FROM DC

In electrical and electronic circuits, it is convenient to think of capacitors as parts which can either block a current or which will permit it to pass. In the circuit of Fig. 6-19, two currents flowing simultaneously through a wire reach capacitor C. One of these currents is dc, while the other is an alternating current. For the direct current the capacitor will behave like an open circuit and so there will be no passage of direct current to the other side of the capacitor. For alternating current, however, the capacitor does not act as an open circuit, but rather as a resistance of some value. The alternating current will then appear on the other side of the capacitor. The amount of current flow may be somewhat reduced, though, due to the opposition of the capacitor. However, what is important here is that the capacitor is being used as a device for separating dc from ac. When used in this connection it is sometimes called a blocking capacitor.

THE CAPACITOR AS A BYPASS UNIT

In still another application, as shown in Fig. 6-20, a wire carrying a direct current and an alternating current is connected to a resistor shunted by a capacitor. The capacitor will block the direct current, but since an additional path is provided, the direct current will flow through the resistor. The alternating current has an option of flowing "through" the capacitor or the resistor. A capacitor used in this kind of circuit arrangement is called a bypass unit, since it permits

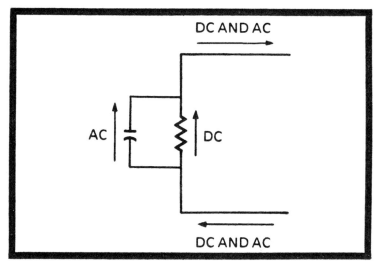

Fig. 6-20. A capacitor is conveniently regarded as a conducting path for ac and an open circuit for dc. In this circuit, the capacitor across resistor helps keep ac from passing through resistor. When used this way the capacitor is called a bypass unit.

the alternating current component of the current to bypass the resistor.

OTHER JOBS FOR CAPACITORS

There seems to be no limit to the number of different jobs capacitors can be made to do in electrical and electronic circuits. They can be used to change the shape of ac waves and to "trigger" other circuits—that is, to start them working or to stop them. They are often used as filters, to smooth small ripples or variations in a current so it can be made to resemble a smooth direct current. As variable units they are used for tuning in radio stations. In short, they can be made to control and guide all the different currents—direct currents and alternating currents—in just about every electrical and electronic circuit. Although basically simple, just a pair of plates and a dielectric, capacitors are every bit as important as resistors and coils.

Tubes and Transistors

While various devices are used to control the flow of current in electric and electronic circuits, none are capable of supplying the fine, instantaneous control possible with tubes and transistors. This does not mean that a current-controlling element, such as a resistor, isn't as necessary as a tube or transistor. It does mean that a resistor has current controlling limitations which are easily overcome by tubes and transistors.

The usual concept of electron movement is the flow of current through some kind of conductor, such as a copper wire. Electric currents, however, not only flow through solids, such as copper, but through gases and also through liquids. Solids, liquids, and gases are often regarded as conducting mediums for electrons, but electrons are quite capable of moving without physical assistance. Electrons can easily travel through a vacuum.

One of the basic problems involving electrons is in getting them to move. A battery is a device for doing this. So is a generator. So are the photocell and thermocouple. In the photocell, electrons are made to move by endowing them with additional energy supplied by light. Such energy can also be furnished by heat.

Fig. 7-1 shows a simple wire structure connected to a battery. Although the wire is in the shape of an inverted V, it can have any other desired shape such as a loop or spiral. If enough current flows through the wire, known as a filament, it will get hot. In an electric light bulb the filament is made hot enough to glow, supplying light. One byproduct of this action is that electrons are literally forced out of the surface of the filament by the intense heat and form a sort of invisible electron cloud around the filament.

Back in 1883 when Thomas A. Edison was experimenting with his early incandescent lamps he placed a metal plate inside one of them and noted that when the plate was made positive with respect to the filament a current flowed from the filament to the plate (Fig. 7-2). When he reversed the connections so the plate was made negative there was no further current flow between filament and plate.

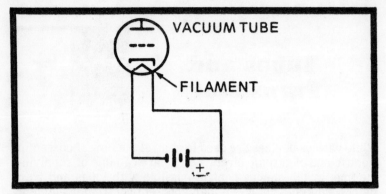

Fig. 7-1. The circle represents a bulb or a tube. When the filament becomes sufficiently hot it emits electrons, which form an invisible cloud around the filament.

Although electrons cannot be seen their movement can be registered by a current reading meter, such as a milliammeter. With the electric light bulb and its metal plate connected as shown in drawing A in Fig. 7-2, the meter gives an indication of current flow. However, with the battery transposed, the meter reads zero, proof that current flow has stopped. This intriguing behavior was called the "Edison effect" but Edison at that time saw no practical application for it. What Edison had done, though, was to invent, unwittingly, the basic vacuum tube—a diode.

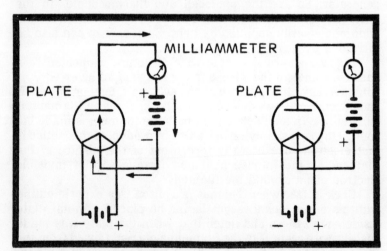

Fig. 7-2. Electrons moving out of the heated filament are attracted to the positive plate (A). When the battery is transposed, as in B, the plate becomes negative and repels electrons.

The Edison effect supplied the groundwork for all future radio tubes. The filament inside Edison's electric light bulb was heated by a battery or a generator. Either could be used since their only function was to supply a current for heating the filament. Because the metal plate inside the electric light bulb was connected to the positive terminal of the battery, it too was positive (another way of saying that it had a shortage of electrons).

Now consider the condition inside the electric lamp. Surrounding the heated filament is a cloud of electrons which have been emitted by it. Nearby is a metallic conductor which has a shortage of electrons. Between the filament and the plate, then, there must be a voltage. A voltage, based on the definition previously supplied, always exists between two substances, generally adjacent, when one has a larger electron supply than the other. Since electrons will always try to remedy such a situation, the electrons now move toward the plate and are unopposed in this movement. Upon reaching the plate, the electrons continue their motion through the meter toward the plus terminal of the battery. The movement is passed along from one electron to the next, through the battery and then along the conducting wire to the filament. As the current flows through the meter, the meter coil turns and in so doing carries the meter pointer along with it. Since the meter pointer faces a scale we can now read the amount of current flow. The current, however, does not stop at the meter, but continues along the connecting conductor to the battery and then to the center point of the filament. The process is a continuous one and as long as the filament is heated and can supply electrons and as long as the battery remains in good working condition, the current will continue to flow.

The battery connected between the plate and the center point of the filament is called the plate battery. It is this battery that supplies the voltage between the plate and the filament and that makes the plate positive with respect to the filament. While the negative lead of the plate battery is shown connected to the filament center, it could more conveniently have been wired to the plus or minus terminal of the filament battery, for this battery is directly connected to the filament.

The arrows in Fig. 7-2A show the path of current flow supplied by the plate battery. Since the current moves in one direction only it is a direct current. The meter which registers the amount of this current is known as a plate meter since it is wired to the plate, and the current it measures is the plate current.

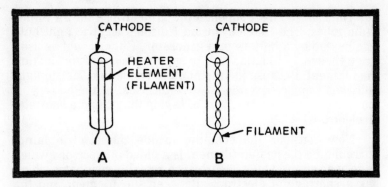

Fig. 7-3. The cathode is a sleeve whose outer surface is coated with an electron-rich substance. The only function of the filament is to make cathode hot enough to emit electrons.

There is still another current in this circuit, although it is not so indicated in Fig. 7-2A. This is the current supplied by the filament battery and used to heat the filament.

Note, in drawing B, the plate battery has been transposed. Since this now makes the plate negative with respect to the filament, electrons boiling out of the filament are repelled. The plate current becomes zero. However, filament current is not affected by this situation and continues to flow. Electrons moving out of the filament remain near it.

Some 21 years after Edison's experiments with the electric lamp, a British physicist, Professor J.A. Fleming, discovered an excellent use for Edison's idea and patented it in 1904 as a two-element diode tube for the detection of radio signals.

FILAMENT AND CATHODE

Fleming's tube is called a diode because it contains two elements or electrodes: di = 2; ode = element; hence, diode. One of the improvements subsequently made in the diode was the design of a cathode as the heated-electron supply source.

As shown in Fig. 7-3, the cathode consists of a sleeve placed over the hot filament. The only purpose of the filament is to heat the cathode and in later years that was actually what the filament was called—heater. The function of the cathode is to emit electrons and so its emitting surface is coated with various chemical oxides capable of doing exactly that. There is no physical connection between the filament and the cathode. In early diodes the filament was heated by a battery. In diodes developed in later years, the filament was connected

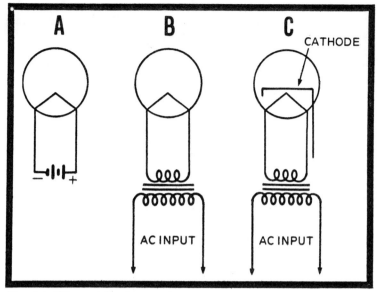

Fig. 7-4. Filament heating methods. (A) Battery heating; (B) transformer heating; (C) indirect heating.

to the secondary winding of a stepdown transformer and was heated by an alternating current (Fig. 7-4).

The diode has a number of useful functions in electricity and electronics, one of which is to help in the conversion of ac to dc. The way this is done is shown in Fig. 7-5. The only change that has now been made is to substitute an ac voltage source for the plate battery. When the top end of the generator

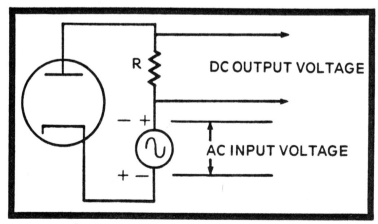

Fig. 7-5. The diode can be used to convert ac to dc.

becomes plus, current flows from the cathode to the anode, through the load resistor, through the ac voltage source, to the cathode, where the entire process starts all over again. However, when the polarity of the generator reverses, the top end of the generator becomes minus. This makes the plate negative and so current stops flowing in the plate circuit. The current flowing through the diode load resistor moves in only one direction. Whenever a current flows through a resistor a voltage is produced or developed across that resistor, in accordance with Ohm's law. Since the current through R is dc, the voltage produced across it is dc. Thus, with the help of the diode, the input voltage is converted into a dc output voltage.

There are two ways in which the current flowing through the diode can be controlled. One of these methods is to increase the voltage across the filament. This raises the temperature of the filament, forcing it to emit more electrons, or, in the case of cathode-type diodes, heating the cathode so it supplies more electrons. The other method is to increase the amount of voltage on the plate by using a plate battery having a higher voltage. Both of these techniques, a hotter filament or cathode, and a higher plate voltage, can be used; but both have limitations. An excessively hot filament will burn out sooner. A higher voltage on the plate means a higher voltage battery, and that can be more costly.

RELATIVE POLARITY

A voltage is electrical pressure and always exists between two points. Every physical voltage source has two terminals: a battery or a generator, with these terminals often identified by plus and minus signs if the voltage source is dc. Ground is often wired to one of the terminals of the voltage source and not only serves as a common connection for other components, but as a reference point as well.

In Fig. 7-6, drawing A, the single battery has a ground connected to its plus terminal. Point A of the battery, then, is negative with respect to the plus terminal but also with respect to ground. However, ground can be connected to either terminal, and so in drawing B, the plus side of the battery, indicated by letter B, is positive with respect to ground. In drawing C there are two batteries, wired in series with the ground connected to their junction. Point A is now negative with respect to ground while point B is positive with respect to ground.

The same situation prevails in drawing D, except that a single battery is used once again. This is the most flexible of the four circuit arrangements, for now point A can be made either positive or negative, or completely neutral with respect to ground, depending entirely on the setting of the slider arm of the variable resistor. When the arm is above center, point A is negative with respect to ground; when below center it is positive; and when at exact center it is neutral.

THE THREE-ELEMENT TUBE

The difficulty with the diode is that while current flows through it can be controlled by changing the amount of anode voltage or by varying the temperature of the filament or cathode, these methods are rather crude. In 1906 Lee de Forest inserted a third element in the tube, placing it in the space between the cathode and the anode. This element was in the form of a wire grid, from which it takes its name—control grid. Fig. 7-6 shows the arrangement of the new form of tube, originally known as an Audion (coined by RCA) but now referred to as a triode because it contains three elements or electrodes.

Part of Fig. 7-7 is familiar, or should be, for the entire arrangement of the diode circuit is still being used. The only new elements that have been added are the control grid and the battery connected to this newly inserted electrode. There are now actually three circuits with the tube as the common connecting link. The first is the filament circuit; the second, the plate or anode circuit; while the remaining circuit is the grid circuit, made up of the cathode, control grid and the battery connected between them. The battery in the grid circuit is called a grid or bias battery.

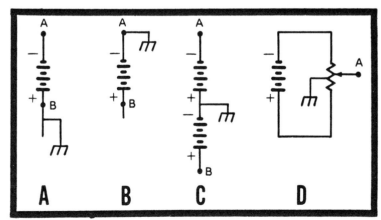

Fig. 7-6. Various ways of indicating relative polarity.

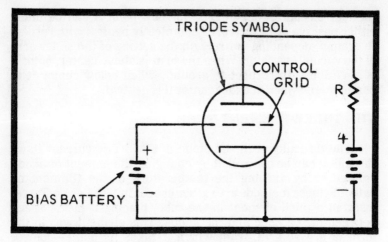

Fig. 7-7. Basic triode vacuum-tube circuit.

Because of the presence of the bias battery, there is now a voltage between the new electrode and the cathode. The control grid is now positive with respect to the cathode, and while the voltage on the grid is very small compared to that on the anode, its influence on the electrons leaving the cathode is quite considerable. Because of the presence of the positive voltage on the control grid, located so much closer to the cathode than the anode, electrons surrounding the cathode are urged forward in the general direction of the anode. By the time they reach the control grid they have considerable velocity. Since the grid is mostly open space, the great majority of electrons rush through and complete their trip to the plate. Some of the electrons do get caught on the control grid but are returned to the cathode for reprocessing, by way of the bias battery.

There are three currents flowing in this triode circuit. One of these is the filament current moving through the secondary of the filament transformer and the filament. This is the heater current and its only function is to heat the cathode, much as a pan on a stove heats the food put in it. The next current is the anode or plate current—the current flowing from the cathode to the plate and then back to the cathode, via the anode battery. Finally, there is a very small current flowing in the grid circuit.

While batteries are shown in the circuits, they can be replaced by electronic power supplies which can furnish dc voltages and currents. When a battery is used in place of the filament transformer it is called an "A" or filament battery.

The battery attached to the anode is the "B" or plate battery, while the battery in the grid circuit is the bias or "C" battery.

The grid is sometimes referred to as a control element, for that is exactly how it behaves. Fig. 7-8 shows a triode tube with a modification made in the grid circuit so that either zero voltage, a positive voltage, or a negative voltage can be put on the control grid.

If the control grid is made sufficiently negative, no current moves in the plate circuit. As the voltage between control grid and cathode is made less and less negative, more and more current passes through the anode circuit. The voltage change on the control grid is very small. This small voltage variation can control a substantial current flow in the anode circuit—a current that can consist of current in the order of milliamperes.

The controlled flow of current in the anode circuit is of no value unless it can be put to work. There are several ways of accomplishing this, one of the simplest being the insertion of a load resistor in series between the anode and the anode battery. Assume this resistor has a value of 50K. Also assume that with the bias control set at its center position, there is zero voltage between control grid and cathode. Under these circuit conditions, the meter connected to the anode reads a current of 1 mA. Since this is the plate or anode current, it also flows through the plate load resistor. However, whenever a current moves through a resistor there is a voltage drop, an IR drop, across that resistor, based directly on Ohm's law. In this

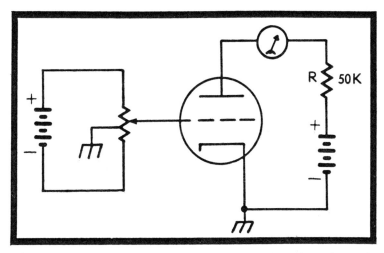

Fig. 7-8. With this circuit arrangement, the control grid can be made either positive or negative with respect to the cathode.

example, the current is 1 mA. To convert milliamperes to amperes, divide by 1000 or move the decimal point to the left three places. Thus:

$$1 \text{ mA} = 0.001\text{A, or } 1 \text{ mA}$$

Using Ohm's law to calculate the IR drop across the resistor:

$$E = I \times R = 0.001 \times 50,000 = 50V$$

Now suppose the bias control is adjusted to put a positive potential on the control grid. This will increase the anode current. If this current increases to 2 mA (and this is entirely possible) the voltage across the load resistor will be:

$$E = I \times R = 0.002 \times 50,000 = 100V$$

The control grid voltage was changed from 0 to +1V, a total voltage change in the control grid circuit of just 1V total. However, the voltage across the load resistor in the anode circuit changed from 50 to 100V, a total voltage change of 50V.

If the voltage on the control grid is made −1V, by moving the control arm of the variable resistor, it is possible that no current at all will flow through the tube. But if there is no current through the load resistor, there can be no voltage across it. The decrease from 0 to −1V in the grid circuit caused a change of 50V across the load resistor in the anode circuit for the voltage across the load resistor dropped from 50V to zero.

AC Input Signal to the Triode

Instead of using the manually operated bias control in the grid circuit, the battery and its shunting resistor can be replaced by an alternator. If the output of the alternator is kept to a maximum of 1V, then the conditions prevailing in the triode circuit will be essentially the same. The output of the alternator will vary between +1V and −1V and so the voltage across the load resistor will vary between 50 and 100V; then 50 and 0V. This new arrangement appears in Fig. 7-9.

A radio signal is also an ac signal and can be substituted for the alternator of Fig. 7-9. Since the voltage variations across the load resistor will faithfully follow the voltage variations of the signal, the voltage across the load will be a replica, a duplicate of the signal. The difference is that the signal voltage (in this example) will have a maximum of 1V while the voltage across the load will have a peak of 50. In effect, then, since the voltage across the load looks like the signal voltage, we have amplified the signal for it is now 50 times as large. The triode now has a signal output that is 50 times as great as the input. This output variation of 50V can be put into another tube and the signal amplified still further.

Diodes and triodes find considerable applications in industrial radio and television circuits. While various other elements have been put into tubes, with some tubes having three or more grids and some having two anodes, the basic operation of the tube is still that of a current controlling device. A tube can be compared to a gun in which a small pressure or force on a trigger releases a much higher force when the gun fires. In fact, a signal voltage is sometimes called a trigger voltage.

When a tube is manufactured, as much air as possible is removed from the tube, so that the inside of the tube is a rather good vacuum, just as in the case of electric light bulbs. Sometimes, though, various kinds of gases, such as argon and neon, are put into tubes for special purposes. These gases supply extra electrons, in addition to those furnished by the cathode, and so such tubes often have very substantial currents.

SOLID-STATE ELECTRONICS

The electrons that flow through a tube move either through a vacuum or through a gas, but it is also possible to have electrons move through a solid substance. This happens every time a current passes through a wire and so getting a current to go through a solid is no problem. The problem in electricity and electronics isn't to get current into motion, but

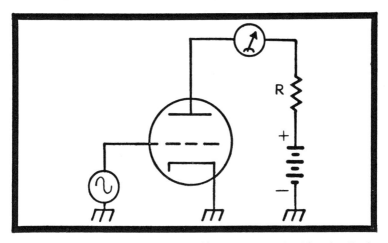

Fig. 7-9. A small ac voltage connected between control grid and cathode can be used to control the current flowing from cathode to plate of the triode. The ground symbol indicates that one terminal of the ac generator, the cathode of the tube, and the negative terminal of the battery are connected.

to get complete, absolute control over that current—to increase it when necessary and to decrease it as desired. The controlled movement of electrons through certain solid substances is known, appropriately enough, as solid-state electronics.

It is very convenient to group all substances into two categories: conductors and insulators. A conductor is a substance that allows the easy passage of electrons through it, while an insulator behaves in an entirely opposite manner, permitting passage of very few electrons. However, it is now possible to form a third group known as semiconductors. These are **not** substances which are sort of halfway between conductors and insulators. Instead, they are basically insulators whose properties have been so modified that a) current can flow through them and b) this current flow can be controlled.

One of these substances is the element germanium. Silicon is another grayish-white metal that is rather brittle. Although germanium is a metal and although we usually regard metals as conductors, in its pure state germanium is an insulator. Germanium is not only a metal but it is a mineral and has a crystalline structure.

The outermost ring of the germanium atom contains four electrons. One germanium atom can establish an atomic union with another germanium atom with each atom sharing its four outermost electrons to form a ring of eight electrons, making pure germanium stable. Since the outermost ring is satisfied, the substance acts like an insulator. No additional electrons are needed.

An element such as antimony has five electrons in its outermost ring. When a small amount of antimony is added to germanium, four of the outer ring electrons of the antimony form a bond with the four outer ring electrons of the germanium. However, the antimony has one electron left over. If a voltage is now applied to this combination of germanium and antimony, a current will flow through it, not as easily as through a conductor such as copper, but much better than through an insulator such as glass. The germanium treated with antimony has been changed into a sort of reluctant current carrier. Germanium treated or doped this way is called a semiconductor and since it now has more electrons than it would have in the untreated state, is known as N-type germanium. The letter N is an abbreviation for negative, since electrons all carry a negative charge.

The antimony used for doping the germanium is a donor because its contribution to the union with germanium has resulted in an excess of electrons. While the antimony has "donated" electrons, it isn't the only substance that can do so. Germanium can also be doped with arsenic, phosphorus, and bismuth. The addition of these elements to germanium does not result in a new chemical compound. The germanium remains germanium and the donor, whichever one is used, remains the same. The real change that has been made is that the germanium now has an electron surplus.

There are various elements that have just three electrons in their outermost orbits. Boron is one of these, aluminum is another, and so are gallium, indium, and thallium. If some of these elements sound strange and rare it is because they are indeed strange and rare. They are sometimes called acceptor elements, since they accept electrons to complete their outer rings.

When one of these substances, such as boron or aluminum, is added to germanium or to silicon the two outer rings of both elements join, but since germanium has 4 outer ring electrons and aluminum or boron have only 3, the result is an outer ring structure of 7 electrons. What the germanium, doped with aluminum or boron, really needs is an additional electron to fill the vacant space. This vacant space can be regarded as a hole into which an electron could fit. Since the outermost ring of the germanium doped with aluminum is short one electron, we can also say it is missing one negative charge. A shortage of electrons, though, is equivalent to calling a substance positive, and so germanium doped in this way is called P-type germanium. The letter P is an abbreviation for positive.

We now have two types of germanium—N-type with an excess of electrons and P-type with an electron shortage.

The Semiconductor Diode

A voltage always exists between two substances when one has more electrons than the other. If a block of N-type ger-

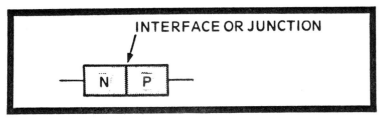

Fig. 7-10. When a block of N-type germanium is put in good contact with a block of P-type germanium, there will be a small movement of electrons from the N to the P block, across the junction.

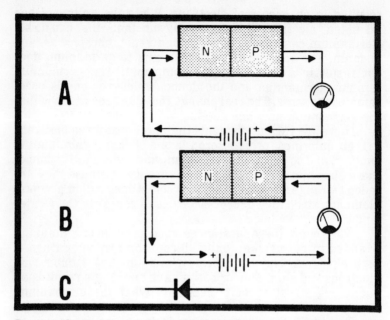

Fig. 7-11. Current flows through the diode when it is forward-biased (A) and practically stops when reverse-biased (B). Drawing C is diode symbol.

manium, as shown in Fig. 7-10, is placed in direct contact with a block of P-type germanium, a very small but still measurable voltage exists between the two blocks. A very limited movement of electrons will take place at the interface or junction of the two blocks with electrons moving from the electron rich N-type over to the electron poor P-type germanium. But as soon as these electrons do cross they will join the outer rings of those atoms of germanium that are at the interface. When they do, those atoms of germanium will now be electrically neutral. They have no further need of electrons nor will they attract any from the N-type germanium. However, behind them, in the remainder of the P-block are germanium atoms that do need and want electrons to complete their outer rings. The only problem is that the satisfied atoms at the interface block any further passage of electrons.

To overcome this situation, Fig. 7-11A shows a battery placed across the joined N and P blocks. Note how the battery is connected. The minus terminal of the battery is wired to the N-block of semiconductor material; the plus terminal is wired to the P-block. The meter in series between the P-block and the plus terminal is used as a current flow indicator. The arrows indicate current flow direction.

More electrons will now crowd into the N-block, pushing their way through the satisfied barrier of germanium atoms in the P-block. So many of them will do this that the P-block will receive more electrons than it needs. The situation in the P-block will be that of a region that is becoming electron saturated. And so electrons will move along the conductor connected to the P-block, over to the plus terminal of the battery where there is actually an electron shortage. At the positive terminal, though, there is a constant migration of electrons over to the negative terminal of the battery because of the chemical action inside the battery. But as the electrons move out of the P-block to the plus terminal of the battery, they are immediately replaced by electrons from the N-block. The N-block, in turn, is on the electron-receiving end since it is connected to the negative, electron-saturated terminal of the battery. And so there is a steady, if small, flow of current through the two doped germanium units. As in the case of the vacuum tube, each of the germanium blocks is regarded as an element, and since there are two elements or electrodes, the unit of Fig. 7-11A is a diode.

Biasing

As long as the battery is connected to the semiconductor diode the meter in the circuit will show a small current. This current is called a forward current and since it moves in one direction only, is a direct current. The method of connecting a battery to a semiconductor diode, as in Fig. 7-11A, is known as forward-biasing.

Fig. 7-11B shows the same germanium diode connected to the battery, but now the battery has been transposed. The plus terminal of the battery is wired to the block of N-type germanium while the minus terminal is connected to the P-block. A few electrons do manage to move from the N-block to the battery and the negative terminal of the battery does supply a small amount of electrons to the P-block, but the current is now very much smaller than in the earlier diode arrangement in drawing A. A diode connected in this way is reverse-biased and the current due to this type of biasing is a reverse current.

The difference between forward and reverse currents is quite substantial. The forward current is in milliamperes and in some cases can be in amperes. The reverse current is typically in milliamperes when the forward current is in amperes and in microamperes when the forward current is in milliamperes. A general rule is that the forward current is 10 to 100 times as great as the reverse current.

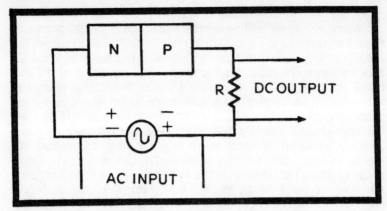

Fig. 7-12. A semiconductor diode can be used as a rectifier—a device for changing an ac input voltage to a dc output voltage.

Drawing C in Fig. 7-11 shows the symbol for the semiconductor diode. The "arrowhead" end is called the cathode while the other end is sometimes referred to as the anode. Electron movement always goes in the direction opposite the arrow in the symbol. This current is the forward current and reverse current would move from the anode to the cathode. The words "cathode" and "anode" are picked up from tube terminology. However, unlike the tube, the cathode of a semiconductor diode does not require heating, while the cathode of a tube has a certain amount of warmup time.

Instead of using a battery, the semiconductor diode can be connected to the output of an ac source. Since the polarity of the source reverses itself regularly, the current flowing through the diode is alternately a forward current and then a reverse current. If the minuscule reverse current is disregarded then we can consider the current flowing through the circuit as a direct current. Thus, the semiconductor diode, as in the case of the vacuum tube diode, works as a device for changing an alternating current to a direct current.

The Diode Rectifier

It's easy enough to put an ac voltage into the diode circuit but to get a dc voltage out of it requires a small circuit modification. Fig. 7-12 shows the arrangement. A load resistor has been placed in series with the P-germanium block. When the polarity of the generator is correct for forward-biasing, a relatively large forward current flows through the diode and, of course, will also flow through the resistor on its way to the battery. With current going through the resistor we have an IR drop across it since all the conditions for the development of a voltage drop are present. We have current or I and we have resistance or R. E = I x R.

Because the current flowing through the load resistor moves in one direction only, the current is dc and so the voltage across the resistor is also dc. A diode working in this way is called a rectifier since it rectifies or changes the ac input voltage to a dc output voltage. There is also the matter of the reverse current, but in a well constructed diode in good condition the reverse current is so very small it can be ignored.

Current Control in the PN Diode

The current flowing through the diode can be controlled in several ways. The generator supplying the ac input could have an attenuator, a device which governs the amount of ac voltage out of the generator. Another method would be to put a variable resistor in series with the generator and to adjust the setting of the resistor until the desired amount of current flows in the diode circuit.

The Transistor Triode

The methods for current control described in the preceding paragraph are satisfactory, but are somewhat

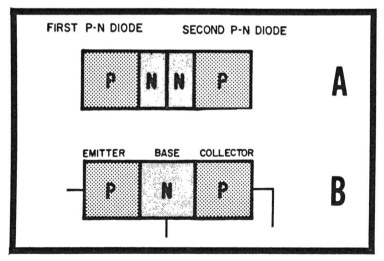

Fig. 7-13. Drawing A shows a pair of semiconductor diodes placed back to back. Drawing B shows the same structure except that the two N blocks have been combined into one.

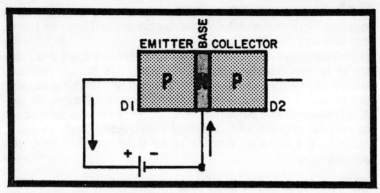

Fig. 7-14. The first diode, D1, is forward-biased. The arrows indicate the direction of the forward current.

limited since they represent manual types of control. A much better control would be one that is purely electronic, just as in the case of the triode vacuum tube.

Fig. 7-13A shows a pair of semiconductor diodes placed back to back, that is, with the N-block of one diode butting up against the N-block of another diode. However, since the N-blocks do touch each other, we can replace them with a single block as in Fig. 7-13B. Since each block is going to be working in a special way, each is given a name of its own. The first P-block at the left is called the emitter, the center N block is the base and the remaining P-block is the collector. This combination of two diodes is called a transistor, with germanium or silicon as the commonly used elements.

Although Fig. 7-13 shows an arrangement of semiconductor diodes so placed that a pair of N-blocks are put against each other, obviously the diodes can be turned around so that a pair of P-blocks are immediately adjacent. A pair of diodes can result in a P-N-P unit or an N-P-N unit. The names given to each block, though, remain the same. The center block, whether P-type material or N-type, is always the base. The block on the left is the emitter and that on the right is the collector.

Steps Toward Current Control

While the two N-blocks have been joined into a single unit, the fact remains that we still have two diodes and we can treat them as such. Thus, in Fig. 7-14 the first diode is forward-biased. The plus terminal of the forward-biasing battery is connected to the emitter and the negative battery terminal is wired to the N-block now identified as the base. At the moment the collector is being completely disregarded. What we have now is a forward-biased diode and a forward current will flow.

The amount of forward current depends on two factors: the way the diode is constructed—that is, how it is doped, and the kind of semiconductor material used. For example, silicon can be used as a semiconductor material instead of germanium, but the behavior of silicon is similar to that of germanium. And, of course, the amount of forward current also depends on the amount of forward-biasing voltage. Current control of the forward current can be had by using the arrangement shown in Fig. 7-15. Here we have a variable resistor shunting a 2V battery and so the voltage on the P-block can be varied from zero to as much as 2V. The higher the voltage on the P-block, the greater will be the current through the emitter and base.

There is an easier way of varying the current through the first diode and that is by putting an ac signal source in series with the forward-biasing battery (Fig. 7-16). Since the polarity of the signal source changes regularly, the effect is to change the forward bias in the same way. As an example, suppose the battery has a potential of 2V and that the ac source an output of 1V. The ac voltage alternately adds to and subtracts from the voltage of the battery. At times, then, the forward-biasing voltage is $2 + 1 = 3V$ and then, when the polarity of the ac voltage changes, it becomes $2 - 1 = 1V$. Consequently, the forward-biasing voltage keeps shifting between 1 and 3V. When the forward bias is 3V the forward current is much larger than when it is just 1V.

The base and collector, representing the second diode, can also be biased by a battery and, as shown in Fig. 7-17A, reverse biasing is used. This means that the current flowing

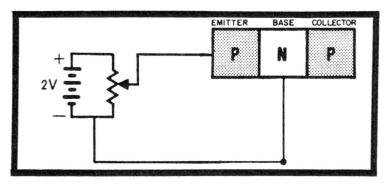

Fig. 7-15. The amount of forward current can be controlled by adjusting the variable resistor. This results in changing the forward-biasing voltage.

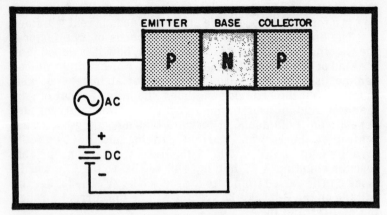

Fig. 7-16. Since the forward bias is now being varied, the forward current will also keep changing in step with the bias.

through the second diode is small compared to that moving through the first diode. And, although the unit in Fig. 7-17A is simply a pair of diodes, it is now supplied with another name and is called a PNP transistor.

In the PNP transistor, the base is the element that is common to the two diodes and in this instance, the base is made of N-type germanium or silicon. However, there is no reason why the diodes cannot be turned around to form an NPN transistor, as shown in Fig. 7-17B. We still have an emitter, base, and collector, except that now the base is P-type material, while the emitter and collector are both N-type. Both diodes are biased—the first diode forward-biased and the second reverse-biased. Compare the connection of the NPN transistor and you will see that by reversing the diodes, the battery connections were also transposed.

The arrows show the paths of current flow in the two transistor types. Current, as usual, flows from minus to plus. In the NPN transistor of Fig. 7-17B battery B1 supplies forward bias to the diode section of the transistor consisting of a N-type emitter and a P-type base. The arrows show that the current moves from the minus terminal of the battery, through the emitter material, through the base material. The arrow pointing downward from the P-type material is the current supplied by B1 and returning to its positive terminal. The first diode is identified as D1.

There is also a small reverse current flowing from B2 through D2. This current flows from the minus terminal of B2, up the connecting wire to the base, through diode D2 and so back to the plus terminal of B2. Since the diodes have been transposed, compared to the NPN arrangement, the batteries have also been transposed, and as a consequence, the movement of current through the PNP transistor is exactly opposite that of the NPN unit.

Transistor Action

To understand how the transistor works, consider the circuit of Fig. 7-18A showing an NPN transistor. The connection to the base has been removed. With the removal of this lead, you can see that B1 and B2 are in series. Electrons move from B1 into the emitter and are promptly attracted to the base because of the electron shortage in the P-type material forming the base. However, from the base there is only one path back to the batteries, and that is through the N-type material of the collector. When electrons move from the base into the collector, other electrons in the collector move out toward the plus terminal of B2. Electrons repel each other and so those electrons in the electron-rich N-type material move

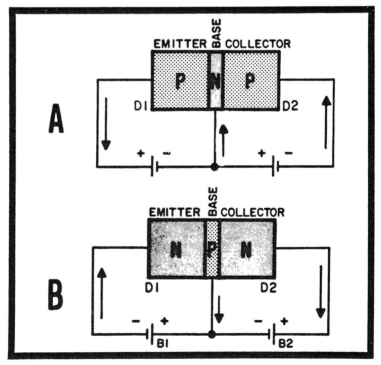

Fig. 7-17. The first and second diodes are both biased. D1 is forward-biased; D2 is reverse-biased.

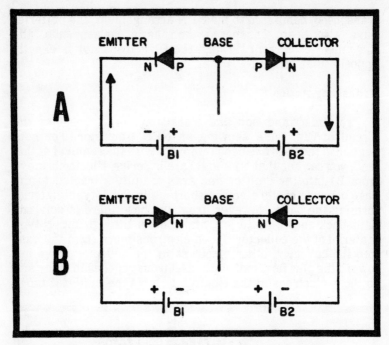

Fig. 7-18. NPN transistor with base lead removed.

out when additional electrons move in. We now have a current flowing through the two diodes, supplied by the two batteries, B1 and B2.

The same results can be achieved by turning the diodes around, provided the connections to the batteries are transposed, as shown in Fig. 7-18B.

The final step toward transistor action involves replacing the connecting wire between the base and the junction of the two batteries, as shown in Fig. 7-19. Two fixed resistors have been added to the circuit: a 300-ohm resistor in series with the wire going to the emitter and a 5.1K resistor in series with the wire attached to the collector.

The current supplied by the two batteries is a total of 5 mA. This is the amount of current which leaves the minus terminal of B1. This current enters the emitter and reaches the base. In the very thin P-region, the current divides. Most of it continues through the N-type material forming the collector, while a very small amount returns to battery B1. In this circuit, assume that 4.7 mA continues through the N-type collector while 0.3 mA leaves the base for the return trip to battery B1.

The current of 5 mA flows through the 300-ohm resistor connected to the N-type emitter. Consequently, there will be a 1.5V drop across this resistor.

Now consider the voltage drop across the 5.1K resistor wired to the collector. The current flowing through this resistor is 4.7 mA; the voltage across it is 0.0047 x 5100 = 23.97V.

As a final step, Fig. 7-19 shows that the circuit includes an ac voltage source having a maximum output of 1V. This ac voltage will alternately add to the voltage of the two batteries and will also subtract or oppose the battery voltage, depending on the polarity of the ac source at any moment. When the voltage of the ac source adds to the battery voltage, the current in the circuit will increase. Assume that it will rise to 7 mA. The voltage across the 300-ohm resistor will go from 1.5V, its original voltage, to 2.1V. The voltage across the 5.1K resistor will increase from 23.97V to 0.007 x 5100 = 35.7V.

When the polarity of the ac voltage reverses and the output of the ac voltage opposes the battery, the current flowing in the circuit will be reduced considerably. But the ratio of the voltage across the 5.1K resistor to that across the

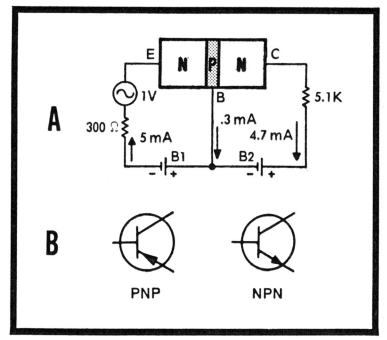

Fig. 7-19. Operating details of an NPN transistor (A); transistor symbols are shown in B.

300-ohm resistor will remain the same. This ratio is 11.73 to 0.6 or 19.5 to 1. What we have here, as in the case of the triode tube, is a small voltage change in one circuit producing a much larger voltage change in another circuit.

Input and Output Circuits

As in the case of the triode tube, the transistor and its various components (resistors and batteries) can be grouped into circuits. The emitter and base, the ac voltage source, the 300-ohm resistor and battery B1 of Fig. 7-19 can be called the input circuit. The base and collector, the 5.1K resistor and battery B2 are referred to as the output circuit.

The ac voltage at the input side of the transistor can be an audio signal supplied by a phonograph or tape recorder, or it can be a radio signal. When the signal voltage increases, the voltage across the output resistor also increases, but much more substantially. The ratio, in Fig. 7-19 is almost 20:1, and so in this circuit, for every 1V change in signal at the input, we get a 20V change in the output. In other words, the effect is equivalent to amplification of the input signal by a factor of 20.

Triode Transistor Symbols

Since there are two basic types of bipolar (two junctions) transistors, NPN and PNP, there are two basic symbols; these are shown in Fig. 7-19B. The symbols are alike except for a difference in the arrow representing the emitter. In the PNP symbol, the arrow points toward the base while in the NPN unit it points away from it.

Waves 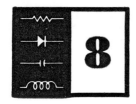 8

The flow of direct current through a component, or the voltage across it, can be pictured in the form shown in Fig. 8-1. This illustration is the graph of current moving through a resistor for a short period of time. In a circuit of this kind the current reaches its maximum value for as long as the switch remains closed. When the switch is opened, the current drops to zero, again almost at once.

A similar graph can be drawn as shown in Fig. 8-2A. However, if the battery connections are transposed, the current flows in a direction that is opposite to its original direction. To show the current flow is in a new direction, the graph is drawn below the horizontal line, as in Fig. 8-2B. The horizontal line, representing units of time in this case, is sometimes called the reference line, or X-axis.

The two graphs can be combined to show the complete action (Fig. 8-3). The reference line indicates that the current has moved in one direction for about 6 seconds (sec). The leads to the battery were then transposed and the current allowed to move in the opposite direction, again for almost 6 sec. At the end of a total of 12 sec, the switch was opened and the current stopped flowing.

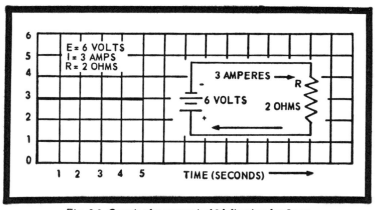

Fig. 8-1. Graph of a current of 3A flowing for 5 sec.

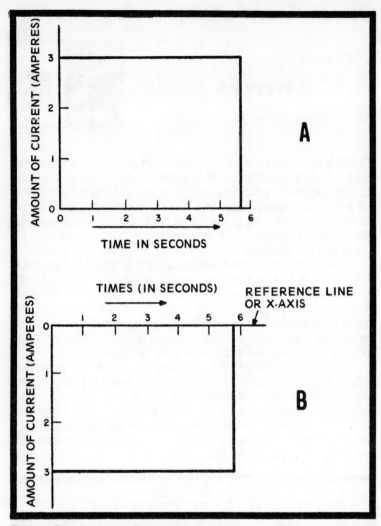

Fig. 8-2. Graphs of the current flowing in the circuit of Fig. 8-1. The only difference between the two is in the direction of current.

Each of the graphs shown in Fig. 8-2A and Fig. 8-2B is called a waveform. Because the waves in Fig. 8-2 look rather square that is exactly what they are called—square waves.

By definition, a direct current is one that always moves in one direction, and an alternating current is one that reverses its direction. The drawing in Fig. 8-2A is that of a direct current for there is no question that the current flows one way only. Similarly, when the connections to the battery are transposed the current is still dc, for although the current now moves in the opposite direction, it does so without reversing itself.

In Fig. 8-4 there is essentially the same circuit but now the switch can be used to transpose the leads to the battery. If the switch is worked back and forth, the current direction will follow the changes of polarity of the battery. Since the current now keeps changing its direction, it is no longer dc but ac. In this case dc has been changed to ac manually, through the use of a hand-operated switch.

The switch in Fig. 8-4 is a double-pole, double-throw (dpdt) type. When the switch contacts terminals 5 and 6, current flows from the minus terminal of the battery, through the switch from contact 3 to 5, down through resistor R, and then from 6 to 4 on the switch, finally returning to the plus terminal of the battery.

When the switch is put in the opposite position, it contacts 1 and 2. Note that 1 is wired to 6 and contact 2 to 5. Current now

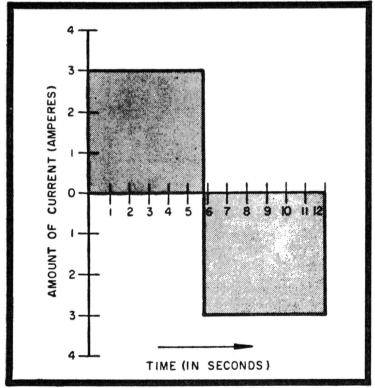

Fig. 8-3. The two graphs of Fig. 8-2 combined.

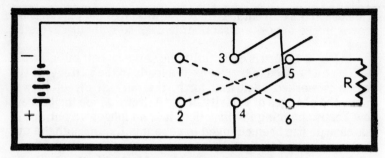

Fig. 8-4. Manual method for changing dc to ac.

flows from the minus terminal of the battery, moving from contact 3 on the switch to 1. From this point the current travels through the wire beneath the switch to contact 6. It then flows up through the resistor and then over to contact 5 of the switch. From here it moves along the wire to contact 2, through the switch blade to contact 4 and then back to the plus terminal of the battery.

THE INVERTER

We now have a back and forth current moving through resistor R. Once again, Ohm's law indicates there is a voltage drop across the resistor since the conditions for obtaining an IR drop are now fulfilled. We have resistance (R) and current (I). $E = I \times R$ and so a voltage appears across the resistor. However, since the current through the resistor is being forced to reverse its direction due to the manipulation of the switch, the voltage across the resistor is ac. Thus, a dc voltage source, plus an additional component such as a switch, can be used to produce ac. A device, such as the switch in this circuit, used for changing dc to ac is called an inverter. The hand-operated switch, though, is just one of many types of inverters.

THE SINE WAVE

Fig. 8-5 is a simplified, stripped-down version of a generator, and consists of an armature coil rotating in the flux field of a magnet. As the armature is turned, an ac voltage is induced across it. This ac voltage causes an alternating current to flow in the turns of wire of the armature coil. The voltage that is produced, however, doesn't have the same waveform as that shown earlier in Fig. 8-1.

Fig. 8-6 is a cross section of the generator with the numbers representing the different positions of the armature.

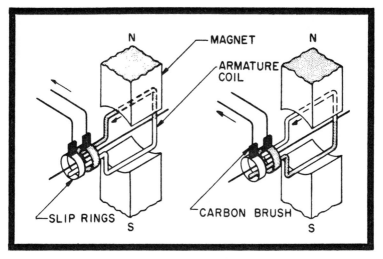

Fig. 8-5. Simplified structure of an ac generator.

Number 1, for example, indicates the armature is horizontal. At 2 the armature has turned 45 degrees in a clockwise direction. At 3 the armature is vertical and has traveled a total of 90 deg.

Toward the right is a graph of the voltage waveform produced by the armature in its circular motion. At 1 the voltage output is zero. As the armature moves in the magnetic

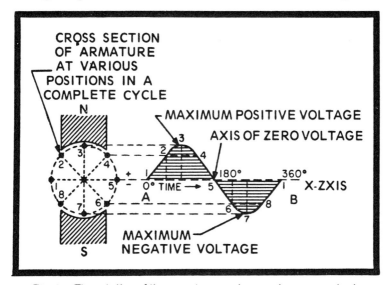

Fig. 8-6. The rotation of the armature produces a sine wave output.

field, its voltage output gradually increases as shown at point 2 on the graph. When the armature is vertical it supplies the maximum output. As the armature passes point 3 its output decreases and at point 5 is zero. During this time the armature has turned 180 degrees or half a cycle. But as the armature continues to rotate, it produces voltage once more, but with reversed polarity. As a result the graph is now shown below the reference line. When the armature finally reaches its starting point the voltage is zero once again. The waveform produced by this generator is called a sine wave.

The fact that part of the wave is above the X-axis and part below it does not mean the current travels like a rollercoaster. The only significance of having the wave above and below the X-axis is to indicate a reversal of polarity and reversal of direction of current flow.

There are many different kinds of ac waveforms, but the sine wave is regarded as the basic ac wave. The maximum part of the wave is called the peak. There are, however, not one but two peaks—a positive peak and a negative peak, since that part of the wave above the X-axis is called positive and that part below it is referred to as negative. The sine wave can be a sine wave of voltage or current.

Some Voltage Differences

If a battery is fresh and in good condition, its output voltage remains constant. An auto battery rated at 12.6V has an output voltage that remains fairly steady at 12.6V. There may be some voltage variations, of course, depending on whether the battery is charging or discharging, the work load and condition of the battery, etc., but these voltage differences are very small.

An ac generator, however, supplies a voltage that starts at zero, reaches a maximum positive peak and then gradually moves toward zero again. The polarity across the terminals of the generator changes and the voltage increases once again, reaches a maximum negative peak, and then gradually decreases to zero again. This series of events keeps repeating, over and over, as long as the generator is in operation. And so, there is quite a difference between the fairly steady, constant output voltage of a battery, and the continuously varying, polarity-changing output voltage of an ac generator.

Because the output of an ac generator doesn't remain still for any length of time, it becomes somewhat difficult to indicate just what the output really is. Part of the time the output is zero. Part of the time, the output approaches maximum. And there are all sorts of voltage values in between. That is why, with an ac generator, or with any device that produces an ac voltage, it is necessary to specify the time or some reference point of measurement. The peak value of an ac wave is one such reference. A generator that has a peak output of 2V is one that will have a positive peak of 2V, followed some short time later by a negative peak of 2V. However, the peak voltage actually represents just a very small fraction of the time the generator is in operation.

Peak and Peak-to-Peak Voltages

The peak voltage of an ac source is its maximum output potential. In electronic formulas it is sometimes written as E_{peak}, or E_p. If the words positive or negative are not used, then it simply means either the maximum positive or the maximum negative voltage. In Fig. 8-7, E_{peak} is 5V.

Sometimes, however, technicians wish to indicate the amount of voltage existing between the positive peak and the negative peak. This is known as the peak-to-peak voltage and is abbreviated as p-p, and in a formula may appear as $E_{p\text{-}p}$. The peak-to-peak voltage in Fig. 8-7 is 10V. The positive peak is 5V and the negative peak is also 5V. The peak-to-peak voltage of a sine wave is equal to 2 x peak, or, in this example,

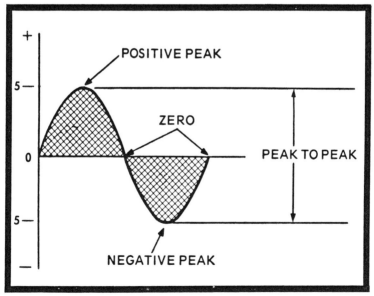

Fig. 8-7. Sine wave peak and peak-to-peak measurements.

235

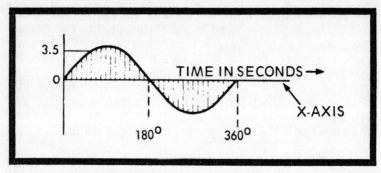

Fig. 8-8. Instantaneous values of a sine wave. Each vertical line represents an instantaneous value.

2 x 5 = 10. This is also applicable to current, and so I_p is peak current and I_{p-p} is peak-to-peak current.

The output voltage of an ac source keeps varying from one moment to the next. The instantaneous value is the value of the wave at any specified moment. In Fig. 8-8 the instantaneous value is indicated by a vertical line drawn from the X-axis until it touches the waveform. Thus, in Fig. 8-8 one instantaneous value indicated by a vertical line is 3.5V. The peak value is also an instantaneous value.

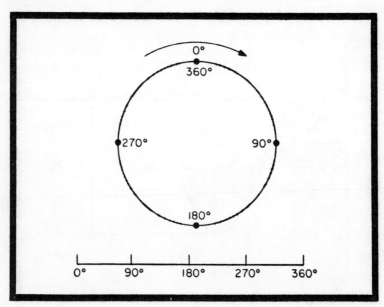

Fig. 8-9. A circle or a straight line can be divided into 360 degrees.

Measurements in Degrees

A circle, as shown in Fig. 8-9, can have its circumference divided into individual units, each of which is one degree. Any circle, regardless of its size, is 360 degrees. If you visualize the circle as constructed of spring metal, it will become flat and straight when snipped. And so a line of any length can also be divided into 360 degrees (deg). The line, of course, can be bent into circular form and the circle would still be 360 deg.

Fig. 8-10 shows a sine wave with the X-axis measured off in degrees. The start of the sine wave is at 0 deg and the finish of the sine wave is at 360 deg. The point at which the waveform crosses the X-axis is 180 deg. The maximum positive peak value, then, is an instantaneous value at 90 deg. The maximum negative or peak value is also an instantaneous value, but is at 270 deg.

Average Value of a Sine Wave

A sine wave can have as many instantaneous values as you want. Each instantaneous value has a certain "height." It is this height, or amplitude, which indicates the value of the sine wave, whether voltage or current, at the particular moment. If all the values of the instantaneous voltages or currents are added, and then divided by the number of instantaneous values selected, the result will be the average value. Assume, for example that 20 instantaneous values are measured for one half of the sine wave. For a wave having a peak of 10V, these instantaneous values might be 1V, 2V, 3V, etc., up to 10V, and then, after the instantaneous peak of 10V

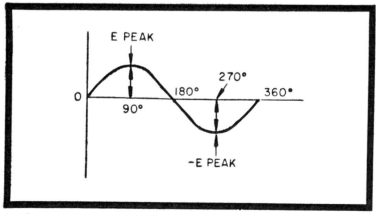

Fig. 8-10. One complete sine wave can be divided into 360 degrees.

is included, could be 9V, 8V, etc. All of these instantaneous values are added and the sum is then divided by 20. (The number 20 is used because a total of 20 instantaneous values were added.) The result is known as the average value of the sine wave. However, it isn't necessary to go through this procedure to find the average value, for through experience in averaging instantaneous values, the average has been found to be 0.637 or 63.7 percent of the peak value. Thus, if the peak value is known (either positive or negative peak, since they are the same) the average value of a sine wave can be calculated by multiplying the peak value by 0.637.

This information can be expressed in terms of a formula:

$$E_{avg} = 0.637 \times E_p$$

Example:

A sine wave has a positive peak of 137V. What is the average value of this wave?

$$E_{avg} = 0.637 \times E_p$$

$$E_{avg} = 0.637 \times 137V = 87.269V$$

Hence, this wave which has a peak of 137V has an average value of only 87.269V.

The same formula can be used to find the peak value of a wave if the average is known, just by making a simple transposition in the formula.

$$E_{avg} = 0.637 \times E_p$$

Divide both sides of this equation by 0.637.

$$\frac{E_{avg}}{0.637} = \frac{0.637 \times E_p}{0.637}$$

Any number or letter divided by itself is equal to 1. 10 divided by 10 = 1. And 0.637 divided by 0.637 equals 1. But that is exactly what we have on the right-hand side of the equation shown above. The equation then becomes:

$$\frac{E_{avg}}{0.637} = 1 \times E_p$$

The number 1 can be dropped because multiplication by 1 doesn't change anything. And so the peak voltage of a sine wave can be found by dividing the average voltage by 0.637.

Example:

A sine wave of current has an average value of 38 mA. What is the peak current value?

$$I_p = \frac{I_{avg}}{0.637} \qquad \frac{E_{avg}}{} = 0.637$$

$$I_p = \frac{38}{0.637} = 59.65 \text{ mA}$$

The significance of this problem is that it indicates a peak current of almost 60 mA, or the maximum current supplied by the ac voltage source. However, part of the time the current is zero. There are also instantaneous values of 1 mA, 2 mA, and so on. The average current is 38 mA.

Effective Value of Sine Wave Voltage or Current

When a direct current flows through a resistor there is a conversion of electrical energy to heat energy. Sometimes this heat is wanted; at other times it is a wasteful byproduct. The heat produced by an electric light bulb is an example of wasted energy. However, heat is deliberately produced in electric devices such as toasters, broilers, and heaters.

Not only a direct current produces heat, but an alternating current also. And so the heating effect of a current supplies a way of making a comparison between direct current and alternating current. The amount of alternating current that will produce the same amount of heat as a direct current is called the effective value or the root-mean-square or rms, value. The rms value of an alternating current can be calculated by multiplying the peak value by 0.707. In terms of a formula:

$$I_{eff} = 0.707 \times I_p$$

Example:

What is the effective current value if the peak current of a sine wave is 838 mA?

$$I_{eff} = 0.707 \times 838 = 592.466 \text{ mA}$$

This formula can not only be used for finding the rms value of a sine wave current, but is applicable to sine wave voltages as well.

Example:

What is the rms voltage of a sine wave if its peak value is 91V?

$$E_{eff} = 0.707 \times 91 = 64.337 \text{V}$$

The formula can be transposed or rearranged to find the peak value of voltage or current if the rms value is known.

$$I_{eff} = 0.707 \times I_p$$

Divide both sides of this equation by 0.707.

$$\frac{I_{eff}}{0.707} = \frac{0.707 \times I_p}{0.707}$$

On the right-hand side of this equation, 0.707 divided by 0.707 is equal to 1. The equation then becomes:

$$\frac{I_{eff}}{0.707} = I_p$$

Example:

A sine wave has an rms current value of 910 uA. What is the peak amount of current flowing in the circuit?

$$I_p = \frac{I_{eff}}{0.707}$$

$$I_p = \frac{910}{0.707} = 1{,}287.1287 \text{ microamperes}$$

This method of finding the peak current when the rms value is known is also applicable to voltage. The formula simply changes to:

$$E_p = \frac{E_{eff}}{0.707}$$

Example:

The rms voltage measured across a resistor through which a sine wave of current is flowing is 423 mV. What is the peak value of the voltage?

$$E_p = \frac{E_{eff}}{0.707} = \frac{423}{0.707} = 598.3027 \text{ mV}$$

In all of the formulas involving average, peak, or rms values of voltage or current, there is no need to convert the units given to basic units. Thus, if the information supplied is in terms of millivolts, the answer will also be in millivolts. If the information is in microamperes, then the answer after the use of the formula will also be in microamperes.

The rms value of a sine wave of voltage or current is determined in somewhat the same manner as finding the average value. First, a selected number of instantaneous values are used. Each of these values is then squared—that is, multiplied by itself. Following the squaring process, all the results are added. The final step is to take the square root of this sum. Fortunately, it isn't necessary to do this work for the effective or rms value turns out to be a little more than 70 percent of the peak value.

WAVELENGTH AND FREQUENCY

A sine wave starts at 0 deg, reaches a peak at 90 deg and then drops to 0 deg again, for a total of 180 deg. This "alternation" of 180 deg is shown in Fig. 8-11. The negative half of the sine wave also requires 180 deg for its completion and is also known as an alternation. One alternation is the positive half of the wave; the other, the negative half. Two such alternations of a sine wave of either voltage or current are called a single complete cycle. A cycle represents one complete action of a sine wave from its start, or 0 deg to its finish, or 360 deg. At 360 deg the wave action repeats and so the second cycle begins. A wave can also be measured in terms of fractions of a cycle. 90 deg represents one-quarter cycle; 180 deg is one-half cycle.

The distance from the start of a single cycle to its completion is called a wavelength, and is usually measured in

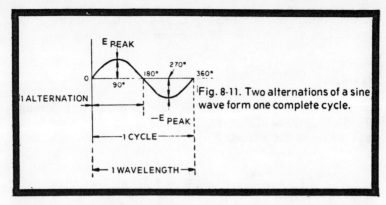

Fig. 8-11. Two alternations of a sine wave form one complete cycle.

meters, or some submultiple of the meter, such as the centimeter. The length of a meter is 39.37 in. or 3.28 ft. 1 meter is the same as 100 centimeters, or, conversely, a centimeter is equal to 0.01 meter. The full wavelength of any sine wave is 360 deg, so the measurement can be made from 0 to 360, or from any other point on the wave to some other point 360 deg away.

The frequency of a wave is the number of complete cycles per unit of time, usually a second. Thus, if 100 complete sine waves of voltage or current are completed in 1 second, the frequency of this wave is 100 cycles per second. If you recall, our term for this is "hertz" (Hz). A sine wave that has a frequency of 500 cps has a frequency of 500 Hz.

Frequency Multiples

The basic unit of frequency is the hertz. A kilohertz (kHz) is 1000 Hz. To convert hertz to kilohertz, divide hertz by 1000. To convert kilohertz to hertz, multiply kilohertz (kHz) by 1000. The easiest way, as in the case of conversions previously described, is to move the decimal point three places to the right or left, as required.

Example:

A sine wave has a frequency of 857 Hz. What is its frequency in terms of kilohertz?

To change hertz to kilohertz, divide by 1000 or move the decimal point three places to the left.

$$857 \text{ Hz} = 0.857 \text{ kHz}$$

Example:

A sine wave has a frequency of 14 kHz. What is its frequency in hertz?

To change kilohertz to hertz, multiply by 1000 (move the decimal point three places to the right).

$$14 \text{ kHz} = 14,000 \text{ Hz}$$

Still another multiple of the hertz is the megahertz or MHz. A megahertz is equal to 1,000,000 hertz. The conversion of hertz to megahertz or megahertz to hertz involves multiplication or division by 1,000,000 or moving the decimal point six places to the right or left as required.

Example:

A sine wave has a frequency of 8,654 Hz. What is its frequency in megahertz?

To convert hertz to megahertz, divide hertz by 1,000,000 (move the decimal point 6 places to the left).

$$8,654 \text{ Hz} = 0.008654 \text{ MHz}$$

Example:

A sine wave has a frequency of 2.4 MHz. What is its equivalent frequency in hertz?

$$2.4 \text{ MHz} = 2,400,000 \text{ Hz}$$

Example:

A sine wave has a frequency of 3,657 Hz. What is its equivalent frequency in megahertz?

$$3,657 \text{ Hz} = 0.003657 \text{ MHz}$$

Sometimes it is necessary to move back and forth between kilohertz and megahertz. The multiplying or dividing factor is 1000 or the decimal point can be moved three places to the right or left, as needed. To convert from kilohertz to megahertz, divide by 1000. To convert from megahertz to kilohertz, multiply by 1000.

Frequency-Wavelength Conversions

Frequency and wavelength are inversely proportional to each other, just another way of saying that as the frequency increases, the wavelength decreases and as frequency decreases, wavelength increases. The formula for converting frequency to wavelength, or wavelength to frequency involves nothing more than dividing the frequency into the speed of light:

$$\text{wavelength (in meters)} = \frac{300{,}000{,}000}{\text{frequency (in hertz)}}$$

In this formula the wavelength is in meters and the frequency is in hertz. Radio waves, like light waves, travel at 300,000,000 meters a second, or approximately 186,000 miles a second in space.

Example:

What is the wavelength of a sine wave whose frequency is 135 kHz?

Since the frequency is in kilohertz while the formula requires the frequency in hertz, the first step is to convert from hertz

$$(135 \text{ kHz} = 135{,}000 \text{ Hz}).$$

$$= \frac{300{,}000{,}000}{135{,}000} = 2{,}222.22 \text{ meters}$$

This means that a wave having a frequency of 135 kHz has a distance of 2,222.22 meters from the start to the finish of each single cycle. Since 1 meter = 3.28 ft, we can convert this into a measurement that is probably more familiar.

$$2{,}222.22 \times 3.28 = 7288.88 \text{ ft}$$

A single cycle of this wave measures over 7000 ft from start to finish.

The same formula can be used for finding the wavelength, if the frequency is known, just by making a transposition in the formula. The formula becomes:

$$\text{frequency} = \frac{300{,}000{,}000}{\text{wavelength}}$$

In this formula, the frequency is in hertz and the wavelength is in meters.

Example:

What is the frequency of a sine wave whose wavelength is 5,400 centimeters?

As a first step, it will be necessary to convert the wavelength from centimeters to meters. A centimeter is equal to 0.01 meter. In this example, the conversion can be done by moving the decimal point two places to the left, equivalent to division by 100.

$$\text{frequency} = \frac{300,000,000}{54} = 5,555,555.55 \text{ hertz}$$

This frequency can be changed to a smaller number, hence a more easily handled number, by converting it to megahertz, moving the decimal left 6 places.

The sum and substance of all this is that a sine wave having a frequency of 5.56 MHz has a wavelength of 54 meters or 54 x 3.28 = 177.12. Not only is the distance from the start to the stop of each cycle of this wave a distance of more than 177 ft, but, according to the frequency, there are more than 5½ million such complete waves each second. The sciences of electricity and electronics supply us with some astonishing facts.

PHASE

It is possible for a conductor, such as a copper wire, to carry any number of currents, alternating currents of different frequencies, or a combination of direct current and one or more alternating currents. It does not follow, though, that all the alternating currents must start and stop their individual cycles at the same time. If, as shown in Fig. 8-12, we have two generators supplying ac voltages, we can start one generator before the other and so the sine waves they produce could be out of step with each other. In the graph of the voltages (or currents) produced by the generators marked A and B, the two voltages are out of step by 90 deg. This separation in degrees, of two voltages, or two currents, or possibly a voltage and a current, is called phase. When voltages or currents start and stop at exactly the same time, they are said to be in phase; if not, they are out of phase. An out-of-phase condition is indicated in degrees.

In the diagram of Fig. 8-13 one of the voltages, E1, has started at zero degrees while the other voltage, E2, shows a beginning at 90 deg. E2 does not start until E1 reaches its peak, at 90 deg. Since E1 starts before E2, we can say that E1 leads E2, or that E2 lags E1.

Sometimes an alternator will contain a single armature coil comprising a large number of turns. There will be just one output voltage from this generator, and so it is sometimes referred to as a single-phase generator, whose output is similar to the single-phase output waveform shown in drawing

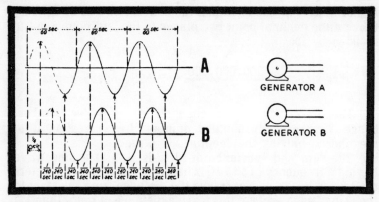

Fig. 8-12. Out-of-phase voltages delivered to a motor. This is a two-phase system (two voltages out of phase). Although two generators are shown in A, it is much more practical to use a single generator with two armatures. The frequency used here is 60 Hz.

A of Fig. 8-12. In some generators there are two armature coils mounted at right angles to each other. This kind of alternator, known as a two-phase generator, supplies two output voltages, out of phase with each other by 90 deg as shown in the waveform in Fig. 8-13. More commonly, a generator will have three separate armature coils, mounted with 120 deg separation. This generator has three output voltages, out of phase with each other by 120 deg as shown in Fig. 8-14. This alternator, a three-phase generator, is much more commonly used than the two-phase type.

A polyphase generator, a generator having two or more independent armatures, can supply two or more voltages which can be used together or independently. Fig. 8-15 shows two generators, although in actual practice both would be combined into one unit. Each armature supplies an ac potential of 115V. Hence a pair of 115V lines would be available from this generator, but these two voltages could be combined into a single 230V line. Since power is the product of voltage and current (P = EI), increasing the voltage reduces the amount of current required for the same power input. 230V is often used as a way of lowering the current requirements for

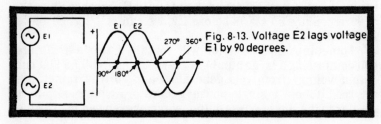

Fig. 8-13. Voltage E2 lags voltage E1 by 90 degrees.

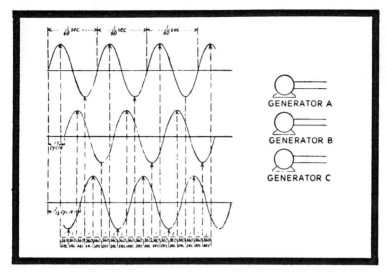

Fig. 8-14. Three-phase system and graph of voltages delivered by the generators. The three generators can be replaced by a single generator having three armatures with each armature separated by 120 degrees. The frequency of this generator is 60 Hz.

devices which normally require large amounts. Thus, electric ranges and large-size air conditioners are often operated from 230V rather than 115V.

While Fig. 8-15 shows a two-phase generator, the same thinking applies to a three-phase generator, except that with such a generator three 115V lines are available.

An alternator is an electromechanical device and so the output frequency is usually rather low. Most commonly, the frequency is 60 Hz or 60 complete cycles per second (Fig. 8-16), although other frequencies, such as 25 Hz, 50 Hz, and 400 Hz are also used.

A generator can be purely electronic and also produce sine wave voltages and currents. Because such generators have no moving parts they can supply frequencies far in excess of those of electromechanical generators. Electronic generators can have sine-wave outputs with frequencies of millions and millions of hertz.

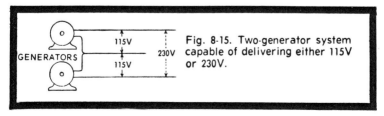

Fig. 8-15. Two-generator system capable of delivering either 115V or 230V.

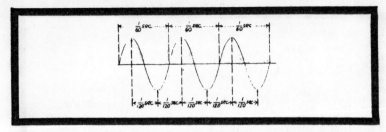

Fig. 8-16. Graph of a 60 Hz ac wave. This consists of 60 complete waves per second. A single complete cycle requires 1/60 second while an alternation is 1/120 second.

WAVEFORM VARIATION

The sine wave is just one of many different kinds of waves. Another type, shown in Fig. 8-17A, is called a sawtooth because of its fancied resemblance to the teeth of a saw. Waves of this kind are used in television receivers and in some test instruments. In a sawtooth wave, the voltage or current increases linearly until it reaches some maximum value, then becomes zero rapidly. This is quite different from a sine wave in which the approach to a peak value is a mirror image of the way in which the voltage or current decreases to zero.

The square wave, mentioned earlier, and now shown again in Fig. 8-17B, reaches a maximum value as quickly as possible, and then, unlike the sine wave or sawtooth, remains at its maximum for most of the time of the wave. The wave then drops to zero as quickly as possible and then, equally rapidly, reaches a maximum negative value which it holds for a part of the time of the wave. Square waves are helpful in making tests of various kinds of electronic amplifiers.

A sound wave, such as the one shown in Fig. 8-17C is a very irregular waveform. It can have sudden starts and stops; can have very high peaks, but may frequently drop to zero. Sound waves are complex waveforms.

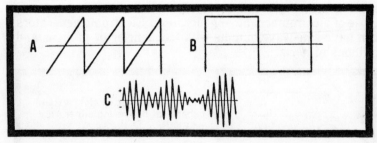

Fig. 8-17. A few of the different types of waveforms: (A) sawtooth; (B) square wave; and C sound wave.

Index

A

A matter of charge	14
Abbreviation	
—microfarad	189
—picofarad	189
Abbreviations	68
AC	71
—and DC combined	75
—generator	147
—input signal to the triode	214
Acceptor	217
Action	
—local	53
—of the capacitor	182
Air-core coil	138
Alternating	
—current	71
—voltage	71
—voltage and current	71, 146
Alternators	149
Aluminum	217
Amber	7, 26
American Wire Gage	122
Ammeters	171
Amount of capacitance	189
Ampere	50
—hour	66
Amperes	62
—milliamperes, microamperes	62
Anode circuit (plate)	212
Antimony	216
Appliances, power ratings	118
Armature	147, 174
Atoms	8, 56
Autotransformer	164
Average value of a sine wave	237
AWG	122

B

Bar magnet	23
Base	222
Basic	
—capacitor	179
—compass	23
—current control	78
—generator	144
—PM motor	24
—resistor	81
Batteries and cells	33
Battery	31
—"duty"	52
—life of a	52
—making connections	53
—the first primary	32
Battery's load	49
Behavior of electrons	10
Bias battery	212
Biasing	219
Blocking capacitor	203
Boron	217
Bound and free electrons	61
Bound electrons	9
Branch currents, how to calculate	94
Brushes	147
"Button" size cells	44
Button-type silver-oxide cells	44
Bypass unit, the capacitors	203

C

Cable and wire	118
Cable, stranded	125
Capacitance	
—decreasing	198
—factors that determine amount of	189
—increasing	197
—stray	199
—units of	188
Capacitor	29
—action of the	182
—as a bypass unit	203
—charging the	180
—codes	188
—leakage	196
—symbols	185
—the basic	179
Capacitors	
—color codes for	193
—electrolytic	190
—fixed	185
—in DC and AC circuits	200
—other jobs for	204
—tantalum	191
—variable	185
Cathode	208
Cell	
—current of a	48
—Daniell	34
—Grove	34
—how it works	34
—internal resistance of	50, 111
—lead-acid	35
—life	38, 42
—primary	41
—putting it to work	35
—the Edison	46
—the mercury	44
—voltage of a	46
Cells	
—and batteries	33
—dry	41
—nickel-cadmium	43
—silver-oxide button-type	44
—types of	34
—wet and dry	39
Centimeter	242
Changing values effect of	106
Charge	
—a matter of	14
—and discharge, controlling	200
—negative	9
—the induced	15
Charges	
—Coulomb's law of	17
—positive and negative	14
Charging circuit	73
Charging the capacitor	180
Chemical	
—energy	135
—method	167
Circuit	
—shunt	93
—symbols	39
—the charging	73
—the series	82
—the series-parallel	101
Circular mils vs square mils (area)	124
Closed circuit	38
Codes and values	81
Coil	
—air-core	138
—and direct current	154
—double-layer	137
—ferrite-core	138
—inductance of	152
—names	137
—single-layer	137
—symbols	138
—triple-layer	137
Coil's magnetic field polarity of	141
Coils in parallel and series-parallel	154
Coils in series	153
Collector	222
Collector rings	147
Color code	
—resistor	108
—transformer	166
Color codes for capacitors	193
Combined AC and DC	75
Commutator	173
Compass, the basic	23
Compounds	8, 59
Concept	
—of inductance	149
—of plus and minus	29
Conductance	97
Conductivity	31
Conductor	
—stranded	128
—the earth as	29
Conductors	60, 216
—the resistance of	30
Connections, making battery	53
Control grid	211
Controlling charge and discharge	200
Conversion rules, current	64
Conversions	82
—frequency-wavelength	244
—voltage	68
Coulomb's law of charges	17
Counter EMF	152

249

Cross-sectional area
of wire ... 121
Current
—alternating ... 71
—and voltage, alternating ... 146
—conversion rules ... 64
—of a cell ... 48
—reading meters, types of ... 171
—short-circuit ... 52
—the direction of ... 70
—the wire of ... 50
Current control
—basic ... 78
—in the PN diode ... 221
—steps toward ... 78, 222
Currents, finding the
branch ... 104
Cycle ... 148

D

D cell, size ... 43
Daniell cell ... 34
d'Arsonval ... 169
DC ... 71
—and AC circuits
capacitors in ... 200
—generator ... 148
DCWV ... 191
Decreasing capacitance ... 198
Degrees, measurements in ... 237
Density, flux ... 20
Depolarizing agent ... 43
Diamagnetic and paramagnetic materials ... 22
Dielectric ... 179, 181
—constant ... 190
Diode rectifier ... 220
Diode, the semiconductor ... 217
Diode tube ... 207
Direct current ... 71
—and the coil ... 154
Direction of current ... 70
Donor ... 216
Double-layer coil ... 137
Double-pole, double-throw
switch ... 130
Double-pole, single-throw
switch ... 130
Dry cells ... 47
—and wet cells ... 39
"Duty," battery ... 52
Dynamic electricity vs
static ... 18
Dynamo ... 148

E

Earth as a conductor ... 29
Edison
—cell ... 46
—effect ... 206
Efficiency, transformer ... 161
Effect
—of changing values ... 106
—of self-induced EMF ... 151
Effective value ... 238
—of sine wave E or I ... 238
Electric lines of force ... 16
Electrical
—energy ... 135
—pressure ... 167
—science, the father of ... 27
Electricity
—how stored ... 179
—magnetism, and electronics ... 24
—the long, long search for ... 26
Electrode ... 32
Electrolyte ... 32
Electrolytic capacitors ... 190
Electromotive force ... 35
Electron
—how it was trapped ... 55

—meet the ... 7
—more about the ... 54
—statistics ... 61
Electronics
—magnetism, and electricity ... 24
—solid-state ... 215
Electrons ... 8
—behavior of ... 10
—bound ... 9
—bound and free ... 61
—free ... 9
Electroscope ... 10, 12
Electrostatic lines ... 11, 17
Elements ... 8
—other ... 59
EMF ... 35, 167
Emitter ... 222
Energy
—chemical ... 135
—electrical ... 135
Expansion method ... 169

F

Factors that determine
amount of capacitance ... 189
Farad ... 189
Father of electrical
science ... 27
Ferrite-core coil ... 138
Field, magnetism ... 20
Filament ... 36, 208
—and cathode ... 208
—battery ... 207
—circuit ... 212
—current ... 207
Finding the branch
currents ... 104
First primary battery ... 32
Fixed and variable
capacitors ... 185
Fixed resistors ... 80
Flashlight cell (No. 6) ... 43
Flux ... 19
Flux density ... 20
Force
—electric lines of ... 16
—electromotive ... 35
—lines of ... 11
Forward
—biasing ... 219
—current ... 219
Free and bound
electrons ... 61
Free electrons ... 9
Frequency
—and wavelength ... 241
—multiples ... 242
—vs wavelength conversions ... 244
Friction ... 27
—machine ... 27
—method ... 168
Frictional electricity ... 168
From a straight wire
into a coil ... 135
From DC to AC ... 134
Fuses ... 132

G

Gage number ... 124
Gallium ... 217
Gauss ... 20
Generator
—polyphase ... 247
—the AC ... 147
—the basic ... 144
—the DC ... 148
Generators vs motors ... 148
Germanium ... 216
"Gimmick" ... 199
Grid circuit ... 212

Ground ... 120
Grove cell ... 34

H

Heat method ... 167
Helix ... 137
Henry ... 152
Hertz ... 148, 242
Horseshoe magnet ... 23
How
—a cell works ... 34
—electricity is stored ... 179
—modern electricity started ... 31
—the electron was trapped ... 55
—to calculate branch currents ... 94
—voltages are generated ... 135

I

Impressed voltage ... 167
Increasing
—capacitance ... 197
—the voltage ... 47
Indium ... 217
Induced
—charge ... 15
—voltage ... 142
Inductance ... 139, 152
—concept of ... 149
—mutual ... 153
—of a coil ... 152
—total ... 153
Inductor ... 139
Input and output circuits ... 228
Input circuits ... 228
Insulators ... 69, 216
Internal resistance
—and the load ... 112
—of a cell ... 50, 111
Inverter ... 232
Ion, negative ... 60
Ionization ... 60
IR drop ... 167
IR drops ... 92

J

Jar, the Leyden ... 28

K

Kilohertz ... 242
Kilohm ... 82
Kilovolt ... 67
Kilowatt ... 117

L

Laminations ... 138
Law
—of charges, Coulomb's ... 17
—Ohm's ... 83
Laws
—of magnetism ... 21
—other power ... 114
Lead-acid ... 45
—cell ... 35
Leyden jar, the ... 28
Life of a
—battery ... 52
—cell ... 38
Light method ... 168
Line current ... 94
Lines, electrostatic ... 11, 17
Lines of force ... 11
—electric ... 16

250

—magnetic 19
Load
—and internal resistance 112
—the battery's 49
Loaded vs unloaded transformers 163
Local action 53
Lodestone 26
—(magnetite) 18
Long, long search for electricity 26

M

Magnet
—bar 23
—horseshoe 23
—ring 23
—temporary 136
Magnetic field 20
—polarity of a coil's 141
—lines of force 19
—materials 21
—poles 19
—poles, similar 21
—poles, unlike 21
—shielding 22
—strength 139
—vs nonmagnetic materials 22
Magnetism
—electricity, and electronics 24
—laws of 21
—the role of 18
Magnetite (lodestone) 18
Magnets, shapes of 23
Making battery connections 53
Manganese dioxide 43
Materials
—diamagnetic and paramagnetic 22
—magnetic 21
Matter, the nature of 56
Maxwell 20
Meaning
—of positive and negative changes 14
—of voltage 66
Measurements in degrees 237
Mechanical method 167
Meet the electron 7
Megahertz 243
Megavolt 67
Megawatt 117
Megohm 82
Mercury cell 44
Meter 242
—sensitivity 171
Meters 169
—current reading types 171
Microammeter 171
Microamperes 62
Microfarad abbreviation 189
Microhenrys 153
Microvolt 67
Microwatt 117
Mil 124
Milliammeter 171
Milliamperes 62
Millihenrys 153
Millivolt 67
Milliwatt 117
Modern electricity, how started 31
Molecules 59
More about the electron 54
Motor, the basic PM 24
Motors 172
—vs generators 148
Multiple transformer 165
Multiples
—frequency 242
—of resistance 81
—of the watt 117
Mutual inductance 153

N

Names, coil 137
Nanofarad 189
Notice of matter 56
Negative 29, 55
—and positive plates 35
—charge 9
—ion 60
Nickel-cadmium cells 43
Nonmagnetic vs magnetic materials 22
Normally closed (N.C.) (relay) 177
Normally-open (N.O.) (relay) 177
North pole (magnetic) 19
NPN transistor 224
N-type 217
—germanium 217
Number of poles 20

O

Ohm's law 83
—and the parallel circuit 95
—and the series circuit 90
—other forms of 85
—simplifying 83
—triangle 88
One-to-one transformer 157
Opposing voltages 72, 151
Other jobs for capacitors 204
Other forms of Ohm's law 85
Other power laws 114
Output circuits 228

P

Padder 185
Parallel circuit 49
—and Ohm's law 95
Parallel, resistors in 96
Paramagnetic and diamagnetic materials 22
Peak and peak-to-peak voltages 235
Percent (resistors) 110
Phase 245
Photocell 168
Photoelectric cell 168
Picofarad abbreviation 189
Plate battery 207
Plate or anode 212
Plus and minus concept of 29
PM motor, the basic 24
PNP transistor 224
Polarity of a coil's magnetic field 141
Polarity, relative 34, 210
Polarization 43
Pole
—north (magnetic) 19
—south (magnetic) 19
Poles
—magnetic 19
—magnetic, similar 21
—number of 20
—unlike magnetic 21
Polyphase generator 247
Positive 29, 55
—and negative charges meaning of 14
—and negative plates 35
Potential 167
—difference 35, 167
Power 113

—in and power out 160
—laws, other 114
—ratings of appliances 118
Primary cell 41
Primary battery, the first 32
Producing voltages 167
P-type 217
Putting the cell to work 35

Q

Quartz 169

R

Ratio, turns 159
Rectifier 221
—the diode 220
Relative polarity 34, 210
Relays 174
Resistance
—and temperature 122
—internal, of a cell 50
—multiples of 81
—of conductors 30
—of resistors in parallel 95
Resistor 78
—color code 108
—the basic 81
—what is it? 78
Resistors, fixed 80
Resistors in parallel 93
—resistance of 95
—three 97
Resistors, variable 107
Reverse bias 219
Reverse current 219
Ring magnet 23
Rings
—collector 147
—slip 147
RMS 238
Rochelle salts 169
Root-mean-square 238
Rotor 185
Rules, current conversion 64

S

Sawtooth 250
Search for electricity the long, long 26
Self-induced EMF effect of 151
Semiconductor diode 217
Semiconductors 216
Separating AC from DC 202
Series circuit 82
—and Ohm's law 90
Series connection 47
Series-parallel
—circuit 101
—coils in parallel and 154
Shapes of magnets 23
Shelf life (battery) 52
Shielding, magnetic 22
Short-circuit current 52
Shunt circuit 93
Silicon 216
Silver-oxide button-type cells 44
Simplifying Ohm's law 83
Sine wave 232
—average value of 237
Single-layer coil 137
Single-pole
—double-throw switch 129, 130
—single-throw switch 129
Ship rings 147

251

Slug (coil)	138
Solenoid	137
Solid-state electronics	215
Some voltage differences	234
Sound wave	250
South pole (magnetic)	19
Square mils vs circular mils (area)	124
Square waves	230
Static electricity	168
Static vs dynamic electricity	18
Statistics, electron	61
Stator	185
Step-down and step-up transformers	159
Steps toward current control	78, 222
Stranded	
—cable	125
—conductor	128
Stray capacitance	199
Strength, magnetic	139
Switches	129
Symbols	
—capacitor	185
—circuit	39
—coil	138
—triode transistors	228

T

Table, wire	120
Tantalum capacitors	191
Temperature and resistance	122
Temporary magnet	136
Thallium	217
The role of magnetism	18
Thermocouple	167
Three-element tube	211
Three resistors in parallel	97
Time constant	200
Tolerance	110
Total inductance	153
Tourmaline	169
Transformer	157
—color code	166
—efficiency	161
—the multiple	165
—the one-to-one	157
Transformers	
—loaded vs unloaded	163
—step-up and step-down	159
Transistor	222
—action	225
—triode	221
Transistors	205
Triangle, the Ohm's law	88
Trimmer	185
Triode	212
—AC input signal to the	214
—the transistor	221
—transistor symbols	228
Triple-layer coil	137
Tube, the three-element	211
Tubes	205
—and transistors	205
Turns ratio	159
Types of	
—cells	34
—coils	136
—current-reading meters	171

U

Unit of current	50
Units of capacitance	188
Units, voltage	67
Unlike magnetic poles	21

V

Values, changing, effect of	106
Values and codes	81
Variable resistors	107
Variations, waveforms	249
Volt	35, 67
Voltage	35
—alternating	71
—conversions	68
—differences, some	234
—drop	167
—increasing the	47
—induced	142
—of a cell	46
—source	167
—the meaning of	66
—units	67
—working	191
Voltages	
—how generated	135
—opposing	72, 151
—peak	235
—peak-to-peak	235
—producing	167
Voltmeters	171

W

Watt	113
—multiples of the	117
Wave, sine	232
Waveform variations	249
Wavelength and frequency	241
Waves	229
Weber	20
Wet cells	45
Wet cells and dry cells	39
What is a resistor?	78
Wire and cable	118
Wire, cross-sectional area of	121
Wire Gage, American	122
Wire table	120
Wire types	125
Working current	52
Working voltage	191

DUE ON ➡ MAR 24, 1979

MATERIALS (EXCEPT FRAMED PRINTS AND TOYS & GAMES) MAY BE RETURNED TO ANY PGCML BRANCH LIBRARY OR BOOKMOBILE.

FINES WILL BE CHARGED ON OVERDUE MATERIALS.

TO CLEAR YOUR RECORD,

THIS DATA PROCESSING CARD MUST BE RETURNED WITH THE MATERIAL. DO NOT REMOVE OR MUTILATE THIS CARD.

PRINCE GEORGE'S COUNTY MARYLAND

IAL LIBRARY SYSTEM, HYATTSVILLE, MARYLAND 20782

BOWIE

7671113
23 BW

537
C
Clifford
Basic electricity & beginning
 electronics.
8.95

BOWIE

AUG 3 0 1976

PRINCE GEORGE'S COUNTY MEMORIAL LIBRARY, MD.